FOOD IRRADIATION

FOOD IRRADIATION

THE MYTH AND THE REALITY

Tony Webb and Tim Lang
of the London Food Commission

THORSONS PUBLISHING GROUP

This revised edition published 1990

First published as *Food Irradiation: The Facts* in 1987

British Library Cataloguing in Publication Data

Webb, Tony, 1945—
Food irradiation. — Rev. ed
1. Food. Irradiation
I. Title II. Lang, Tim
664' .0288

ISBN 0-7225-2224-X

Published by Thorsons Publishers Limited, Wellingborough, Northamptonshire NN8 2RQ, England

Printed in Great Britain by Mackays of Chatham, Kent

10 9 8 7 6 5 4 3 2 1

Making sense of it all

Sometimes in struggling to make sense of a mass of conflicting evidence and argument one stumbles across a simple question that makes sense of it all.

It took three years of research and campaigning before the obvious became clear to us. Put simply: if any food had to be irradiated, consumers will need to ask, what was wrong with it? Good food doesn't need irradiation.

List of tables and figures

About the authors

Tony Webb MSc has been monitoring developments in food irradiation world-wide for the past five years, and has been actively involved in researching and writing about radiation and health issues for the past ten years.

Tim Lang PhD has been director of the London Food Commission since 1984 and is the author of numerous publications on food policy issues. He has been instrumental in putting food and food policy issues on the UK public agenda.

They have collaborated on this project for the past five years.

Contents

Preface

This book has grown out of our work at the London Food Commission. It is a new, revised and updated version of work some of which has appeared in several earlier publications:

Food Irradiation — The Facts, 1987.
Food Irradiation — Who Wants It? written with Kitty Tucker in the USA, 1987.
Food Adulteration and How to Beat It written with other members of the Food Commission staff, 1988.
Food Irradiation in Australia, written with Beverly Sutherland Smith, 1988.
Numerous briefing papers, articles, submissions to government/parliamentary inquiries, conferences etc.

It seeks to present the key issues which need to be resolved before the controversial food processing technology of irradiation is permitted.

Five years ago, when we began our investigations, we were critical but not opposed to irradiation. Far from diminishing, our critique has grown as we have uncovered fraud and abuse, lies, distortion, misuse and misrepresentation of some of the scientific evidence and a mass of wishful opinion masquerading as fact. Irradiation is being promoted with scant regard for the facts — by myth rather than reality. The benefits of irradiation are largely illusory. The risks are very real. In these circumstances we find it necessary to declare ourselves reluctant opponents of irradiation.

Our concerns are not limited to the UK. Irradiation is being promoted internationally. What we have uncovered is a global plan for gaining acceptance of irradiation over the heads of consumers who neither need nor want it. We have been active in helping to build an international network of organizations and individuals who, like ourselves, oppose this global marketing plan. The network spans rich and poor, developed and developing countries.

The book is written from these perspectives. We believe that the facts speak for themselves.

Acknowledgements

No book is ever written by its authors alone and this has been no exception. Those who have contributed are too numerous to name individually but special thanks and acknowledgement must go to:

Kitty Tucker who co-authored the US book and contributed valuable research material to it

Frank Cook MP and members of the London Food Commission Working Group on Food Irradiation

Gerd Billen-Girmscheid, Jesper Toft and other members of the European Food Irradiation Network

Anwar Fazal, Martin Abraham and staff and member organizations of the International Organization of Consumers Unions

Dr Vijayalaxmi

Food Chain Coalition (Canada); National Coalition to Stop Food Irradiation, Food and Water Inc., Michael Jacobson, Centre For Science in the Public Interest and Health and Energy Institute (USA); Mark Lawrence, Food Industries Unions Federation, John Scott MP, Beverly Sutherland-Smith and National Coalition to Stop Food Irradiation (Australia); Consumers' Union of Japan; Voluntary Consumers Group of Thailand, Education and Research Association for Consumers (Malaysia); Environmental Liaison Centre International (Nairobi, Kenya); International Union of Foodworkers; and many other consumer, environmental, trade union and public health agencies we have worked with around the world

Ken Bell MBE, who blew the whistle on abuse of irradiation and supported research undertaken by the North London Polytechnic and the London Food Commission

Sue Dibb, Liz Castledine and staff of the London Food Commission

Leah Bloomfield for the framework for analysis of the safety of irradiated food and for many helpful suggestions during the writing of this edition

The staff at Thorsons Publishing Group.

Tony Webb and Tim Lang
London Food Commission
October 1989

1. The best thing since sliced bread?

Changing priorities — the global food economy

We all need to eat. We all want food which is safe, wholesome and nutritious. In the richer nations, people can also expect to have most of what they want when they want it. They can have fresh fruit in winter, seafood in the middle of a continent and exotic foods from just about anywhere on the globe. They frequently want some of this food processed and packaged for convenience so that less time has to be spent preparing meals. They are also demanding food with fewer additives, pesticide residues and other forms of adulteration. Food quality is a consumer issue and the food industry is having to respond to consumer-led demand.

On the other hand, the food processing industries excel in both creating and satisfying the demand for variety and convenience. Processed food is simple to prepare and cook. It can also be made to keep longer by cooking, salting, drying, bottling, canning, packaging, smoking, chilling, freezing, dehydrating, and using chemical additives. As a result, the developed world now enjoys, literally, the fruits of the earth, and access to just about every food available anywhere on the globe.

There is, however, hunger and malnutrition amidst plenty. Large sections of the population of even the most developed countries have dietary deficiencies. United Nations agencies and national governments are now recommending dietary change as a way of promoting public health and reducing the risk of coronary heart disease and cancers. People are encouraged to eat less fat, salt, sugar, refined starch and processed food, and eat more white meats, whole grains, fresh fruit and vegetables. All of this is changing the face of the food industry and the expectations we have of food and a balanced diet. Already we can see this new balance between the priorities of sufficiency, variety, quality and convenience being reflected in the policy of leading supermarket chains. These emphasize freshness, variety and, in many cases, insist their suppliers provide additive-free alternatives to many common processed goods.

We are also aware of world hunger. The support given to famine-aid appeals indicates that there is a sincere and genuine concern for those who do not have enough food or enough of the right kind of food. There is also a growing awareness of the economic trap faced by many countries, which need to grow 'cash' crops for export to earn foreign currency while, at the same time, they face food shortages and malnutrition at home.

The global food economy has failed to share the benefits of increased production and international trade with the less developed countries. Apart from extreme cases of famine and drought, the problem of hunger and malnutrition is not one of shortage but of failure to distribute the food to those in need. Much of this is undoubtedly a result of economic factors. The rich can afford the food; the poor cannot. However, in many areas, some 25 per cent to 30 per cent of the food is wasted for lack of transport and storage. Any additional technology that can preserve food deserves consideration.

But each preservation technology has a price — both in money terms and in terms of the damage it does to the quality of the food. Processing and storage inevitably result in loss of essential nutrients. Various additives which were believed to be safe have later been shown to be hazardous. Freezing may damage the texture of foods. Even cooking causes some effects that are undesirable from a health standpoint. There is no system for preserving food that is 100 per cent perfect.

Food irradiation

It is in this context that we consider irradiation — the latest in this long line of food processing technologies. Irradiation involves using very large doses of ionizing radiation which, it is claimed, will preserve and improve the safety of food. It has been hailed as an alternative to other methods of preservation such as use of chemical additives. It is claimed that the process is completely safe, and that consumers will benefit from reduced wastage, greater convenience and better quality food.

We disagree. This book will investigate the pros and cons of irradiation. It will illustrate our growing concern and disillusionment with this controversial technology. Five years ago we were critical but not opposed to irradiation. Today, we find almost nothing to commend it. There are serious questions about its safety and about the effects it has on food quality. There are also serious questions about the objectivity and integrity of those in the food, nuclear and irradiation industries who promote food irradiation — questions that need to be put to those in government and science who have decided to approve its use.

To its promoters, irradiation is the best thing since sliced white bread. It is a process which is completely safe and beneficial. It can extend the shelf-life of food, reduce food wastage, make food safer by killing bacteria that could cause food poisoning and sometimes substitute for other, possibly hazardous, methods of food preservation. To its critics, irradiation has yet to pass the kind of safety-testing programme that would inspire confidence. It clearly reduces the nutritional value of food. It is a technology that can be, and is being, abused by food and irradiation companies to conceal unhygienic processing and handling and so put unsaleable food back on the market. It is a technology that, furthermore, makes obsolete many of the current controls by which public health agencies can prevent this happening. It allows old food to be sold as fresh, and contaminated food to be sold as clean. In short, it is a technical fix which covers up low-quality, stale, dirty food and allows the unscrupulous to avoid the need to tackle problems of quality and hygiene at the source.

In this book we will chart the examples of fraud and bad science that characterize the investigation of the safety of irradiated food. Recent history is littered with examples of materials and processes being declared safe only to be found to be highly dangerous. The lessons of thalidomide, asbestos, and a host of industrial chemicals should have been learned before now. The public has every reason to demand a stricter investigation of safety of any new technology. Yet the assurances on the safety of food irradiation being offered by some governments, some scientific bodies, some sections of United Nations agencies and, of course, those who stand to benefit from irradiation are based on selective, biased, illogical, misleading and inaccurate use of scientific evidence. These assurances of safety are inadequate. Is it that the promoters of food irradiation are unable to make seemingly obvious connections between related evidence suggesting possible health risks, or has there been a deliberate attempt to cover it up? Is this simply incompetence or is it deliberate scientific fraud?

If it had to stand on its merits, irradiation would simply be ignored by most of the food industry and the consuming public. There is simply no demand for it. Unfortunately, the world is not merely being offered irradiation on a 'take it or leave it' basis. It is being actively promoted on the basis that it is needed to deal with major issues of public concern — food poisoning and world hunger. Far from resolving these problems, irradiation may actually make them worse unless a wide range of other measures are taken. If these other measures are taken, irradiation becomes largely irrelevant. In short, it is hard to escape the conclusion that irradiation is a technology looking for a use. It is being promoted in a less than impartial manner by national and international agencies with hidden agendas and by companies with vested interests. The promotion campaign is using expert opinions and emotive arguments rather than facts. It is doing so in the face of world-wide, rational and informed opposition from consumer, environmental, medical, public health, worker and food industry organizations.

We will also suggest what you, the reader, can do about it. In the UK, the government has decided to remove the ban on irradiation that has been in place since 1967. The European Commission has proposed uniform legislation that would require all European Community countries to permit free trade in irradiated foods by 1992. Considerable efforts are being made by the International Atomic Energy Agency to promote irradiation in 'Third World' countries. Several countries in Europe already permit irradiation and some actually use it for selected foods. A similar situation exists in the USA, Canada, South Africa, Israel, China and several Eastern European countries.

Despite this level of acceptance and continuing promotion,

irradiation of our food supply is not inevitable. It is worth remembering that, of some 160 nations world-wide, only 35 have given permits for irradiation and fewer still actually use it. It is still a technology in its infancy. A number of countries such as West Germany, Sweden, Australia and New Zealand do not permit irradiation at all. Several others allow only extremely limited use, for example, spices in Denmark and potatoes in Japan. There is strong consumer and government resistance to any further relaxation of controls in these countries. Even in countries which permit irradiation, there is still only very limited use. The total capacity of all irradiation facilities in the European Economic Community is only 47,500 tons per year. Much of this is used for irradiation of medical and surgical supplies. Even if all these facilities were used solely for food, the amount would be trivial compared to the total volume of food traded within the EEC.

In the USA, where irradiation of pork, grains, spices, fruit and vegetables is permitted, it appears that only spices are actually being irradiated and consumers are not told when these are used. There is no honestly labelled irradiated food on sale. The few attempts to test-market irradiated foods in the USA met with such resistance that the stores stopped the trials and several announced they will not sell irradiated foods in the future. If the USA complies with the 1989 United Nations CODEX Committee agreement that all irradiated foods, including ingredients, should be clearly labelled, the use of irradiated spices will face similar consumer resistance.

Most food manufacturers are unwilling to label irradiated foods and face consumer rejection of their product. Use of irradiation is a clear warning sign that all is not well with some foods. Indeed, the key question which underlies our concern throughout this book is, **if any food had to be irradiated, what was wrong with it?**

If there was nothing wrong, it certainly didn't need irradiation. It is bizarre indeed that spices, long associated with food-preservation qualities, should now need irradiation to reduce the risk of contaminating other food with poisonous bacteria. Something is clearly very wrong with the way some spices are now produced and processed.

The image of some seafood, such as shrimp and prawns, has already been damaged as a result of scandals where food companies have used irradiation to hide unacceptable levels of bacterial contamination on unsaleable food. Good, clean food doesn't need irradiation. There are countries, particularly in South and South-east Asia, which are making strenuous efforts to improve hygiene, and there are companies in the international seafood trade that have never needed irradiation and who have declared that they will not use it. The director of one of the largest of these companies has taken the unusual step of 'blowing the whistle' on the abuse of irradiation he knows is going on within the seafood trade.

Other sections of the food trade are also aware of the dangers that irradiation poses to the image of their product. It will certainly not help the poultry industry to tell consumers honestly why irradiation has been used — that processing has burst the chicken gut, spread faeces over the carcass, failed to clean it off thoroughly and that irradiation has been used to hide the resulting contamination with *Salmonella* food-poisoning bacteria. An industry that suffered a 30 per cent fall in egg sales in the UK in 1988/9 as a result of a public health scare over *Salmonella* has every reason to avoid irradiation.

The food industry faces a clear choice: to improve hygiene in the processing of food, or to use irradiation to hide the contamination. The consumer faces a similar clear choice: to buy clean food, thereby encouraging the companies that pay attention to hygiene, or to accept irradiated food and legitimize the worst hygiene practices.

In many countries the key to acceptance of irradiated foods will be the large supermarket chains which dominate the retail trade. In the UK three of the large retailers have publicly declared that irradiation does not fit in with their policy of providing good quality, fresh food. A major campaign is underway in the USA to obtain similar positive statements from thc large supermarkets.

Finally, in democratic countries, governments exist to implement the will of the people. This may seem to be a simple, almost naive statement, but a brief review of the way that governments have responded, albeit slowly, to the growing concerns of consumers over food issues such as additives, pesticide residues, labelling, water quality, and bacterial food poisoning shows that therc is considerable scope for influence leading to legislation and control.

We hope that, armed with the information in this book and the suggestions for action, you will be able to play a part in influencing the outcome of the campaign for food quality by making irradiation what it should be — irrelevant.

2. What is food irradiation?

Put simply, food irradiation is a treatment involving very large doses of ionizing radiation to produce changes in food which are considered desirable. These changes will allow the food to be stored longer by delaying ripening or sprouting, and will kill some insects and bacteria.[1] In this chapter we will explore how this is done and outline the history of its promotion as a technology for preserving and improving the safety of food. In later chapters we will look more closely into its safety, whether it can deliver all that is promised of it, and who is behind the promotion of this controversial technology.

Ionizing radiation

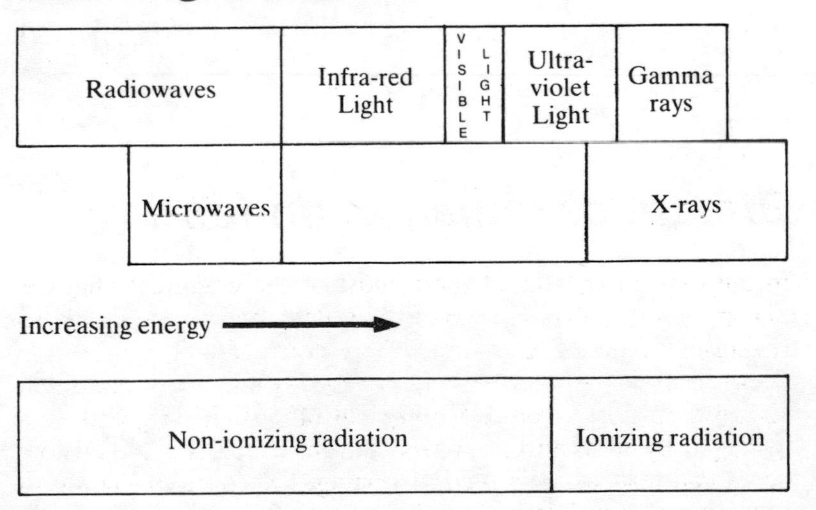

Figure 1: Radiation

Radiation is a household word for a wide spectrum of electromagnetic energy. At the low end of the spectrum are the emissions from power lines and a wide range of electrical equipment including computer terminals and household appliances. Higher up are radiowaves and microwaves. Also included are infra-red, visible and ultraviolet light, and at the upper end are found X-rays and the gamma rays given off by radioactive materials.

All matter is made up of atoms and molecules. An atom can be regarded as consisting of a central nucleus with a cloud of electrons orbiting around it. Molecules are combinations of various atoms, the atomic type and arrangement determining the nature of the material. When radiation strikes a material it transfers some of its energy. This energy transfer can cause the atoms and molecules to vibrate more rapidly. This results in heating, for example, of matter lying in the sunshine or cooking with a microwave oven. At a ccrtain point on the energy spectrum, however, the radiation has sufficient energy to alter the atomic structure of the material it strikes. Molecules can break apart and recombine to form new and different materials. The process involves knocking electrons out of their orbit around the nucleus of the atom, leaving positively and negatively charged particles called 'ions'. Above this energy level the radiation is called '*ionizing*' radiation. The ions and what are called 'free radicals' — uncharged, highly reactive, single atoms or parts of molecules — are chemically very active. They can initiate a wide range of chemical reactions which can alter the chemical structure of the irradiated material.

Effects of radiation on food

In the case of irradiated food, most of the chemical changes occur very rapidly. However, it is possible to set up chain reactions, especially in some fatty constituents, which can continue for several weeks after irradiation.[2] The chemical changes, in turn, produce biological effects. Living cells are damaged by irradiation. High doses invariably kill. Even low doses can lead to cells malfunctioning. Exposure of people to low doses of radiation has been linked to cancer, genetic defects and other health defects.[3] [4] [5] Insects and parasites can be killed or rendered infertile by the massive doses used in food irradiation. Some bacteria can be killed, as can some of the more sensitive food cells such as the sprouting tip of onions or potatoes — thereby preventing them from sprouting. Other effects include destruction of the enzymes that cause fruits to ripen and, eventually, decay — thereby extending the 'shelf-life' of these foods.[6]

Not all the effects are beneficial or desirable. Irradiation also softens the tissues of many foods. Meats bleed more; fruit and vegetables are softened, leaving them more vulnerable to bruising and attack by rot and mould. Some of the chemical changes produce undesirable flavours and odours.[7] Other foods also develop 'off' flavours and tastes. The art of irradiation is to be able to achieve the desired effects without the undesirable ones.

Irradiation of food is clearly different from use of microwave radiation which heats but does not rearrange the atomic structure of the food's components. It is also very different from radioactive contamination of food, such as occurred in many parts of Europe after the accident at the Chernobyl nuclear plant in the USSR. In this case, radioactive material was dispersed into the atmosphere and was carried over large areas. The fallout of this radioactive debris contaminated soils, was taken up by plants, consumed by animals and thus entered the human food supply. In irradiation of food, provided there are no unforeseen accidents, no such contamination occurs. It is like the difference between having an X-ray and having a medical examination which involves inserting some radioactive material into the body. With an X-ray, the patient does not leave the hospital radioactive. The rays pass through the body from the X-ray machine and the shadow of bones and organs is recorded on a photographic plate and that is it! On the other hand, some 'nuclear medicine' procedures involve giving small amounts of radioactive material to patients. Having a radioactive implant inserted into a tumour to kill cancer cells results in the patient being made radioactive for a short time — until the isotopes used are withdrawn, excreted or their radioactivity dies away.

With food irradiation, the radioactive sources are sealed so as to prevent contamination of the food. Radioactive bombardment of some materials can cause secondary radioactivity to be created. The amount and type of this induced radioactivity depends on the energy of the source radiation. For this reason, various national and international scientific bodies have recommended that only low-energy ionizing radiation be used on food.[8] The radioactive sources are carefully chosen so that they do not cause significant amounts of radioactivity to be created in the food.

Sources of ionizing radiation for food irradiation

Ionizing radiation for food irradiation can be provided by:

- *electrical energy/machine sources.* Electric and magnetic fields are used to accelerate electrons to high speeds and these can be used to bombard the food. Electrons, however, have relatively low penetrating power and this approach is only useful for very thin packages or where only the surface of the food needs to be irradiated. Alternatively, the accelerated electrons can be made to strike a metal target and so be converted into X-rays — much like the equipment used in hospitals and dental surgeries, but larger. These X-rays are more penetrating and can be used on relatively large food packages.
- *radioactive materials.* Two particular materials have been found to be appropriate for irradiation of food — cobalt 60 and caesium 137. These have been chosen because the energy level of the gamma radiation they emit is relatively low in the ionizing part of the spectrum. Caesium and cobalt are both by-products of nuclear reactor technology.

Caesium 137 is created as a fission product in the spent fuel from a nuclear reactor. It can be a problem as part of the nuclear waste because it has a long half-life. It takes 30 years for half the material to undergo radioactive decay. It will be significantly radioactive, and hence a disposal problem, for 2–300 years. The caesium can, however, be extracted during what is called 're-processing' — the chemical separation of re-useable elements in spent fuel. This is currently done mainly to extract plutonium for the fast-breeder reactor programme and for nuclear weapons. Using caesium for food irradiation is a bonus for the nuclear industries as it creates a market for an otherwise troublesome waste material.

Cobalt 60 is artificially created from non-radioactive cobalt 59 in some types of nuclear reactor. It has achieved widespread use as a medical and industrial isotope for weld and metal fatigue testing, and treatment of cancer patients in hospitals. Cobalt 60 is also produced as a waste product from the steam generators of the 'pressurized water' nuclear reactors. With a radioactive half-life of five years this can present a disposal problem for some 30-50 years. So far, however, there is no commercial extraction of this waste cobalt for use in food or other irradiation processes.

Food irradiation facilities

Irradiation facilities have existed for a number of years for the purpose of sterilizing medical supplies and equipment such as gloves and syringes. Some companies such as Johnson and Johnson have used irradiation on tampons and sanitary pads. The facilities for irradiating food are merely an extension of this technology. Sometimes food, medical supplies, and such products as wool, carpets or plastic-coated wire can be irradiated in the same facility. This is regarded by the various international promoters of irradiation as acceptable, though irradiators of food in both South Africa and Israel have expressed some reservations about the wisdom of this mixing of products on technical and health grounds.[9]

In general an irradiation facility consists of:

- *the irradiation cell* where food is exposed to the radioactive material or machine source. Radioactive material sources will usually be stored in a water-filled pool below the cell when not in use. Machine sources can simply be switched off. Thick concrete shielding protects workers from direct exposure to the source.
- *a loading facility* where the food to be irradiated is packaged and pre-treated by heating and/or refrigeration as needed, and loaded onto conveyors that will carry it into the irradiation cell.
- *a storage facility* where it is removed from the conveyor after irradiation and stored at the required (usually low)

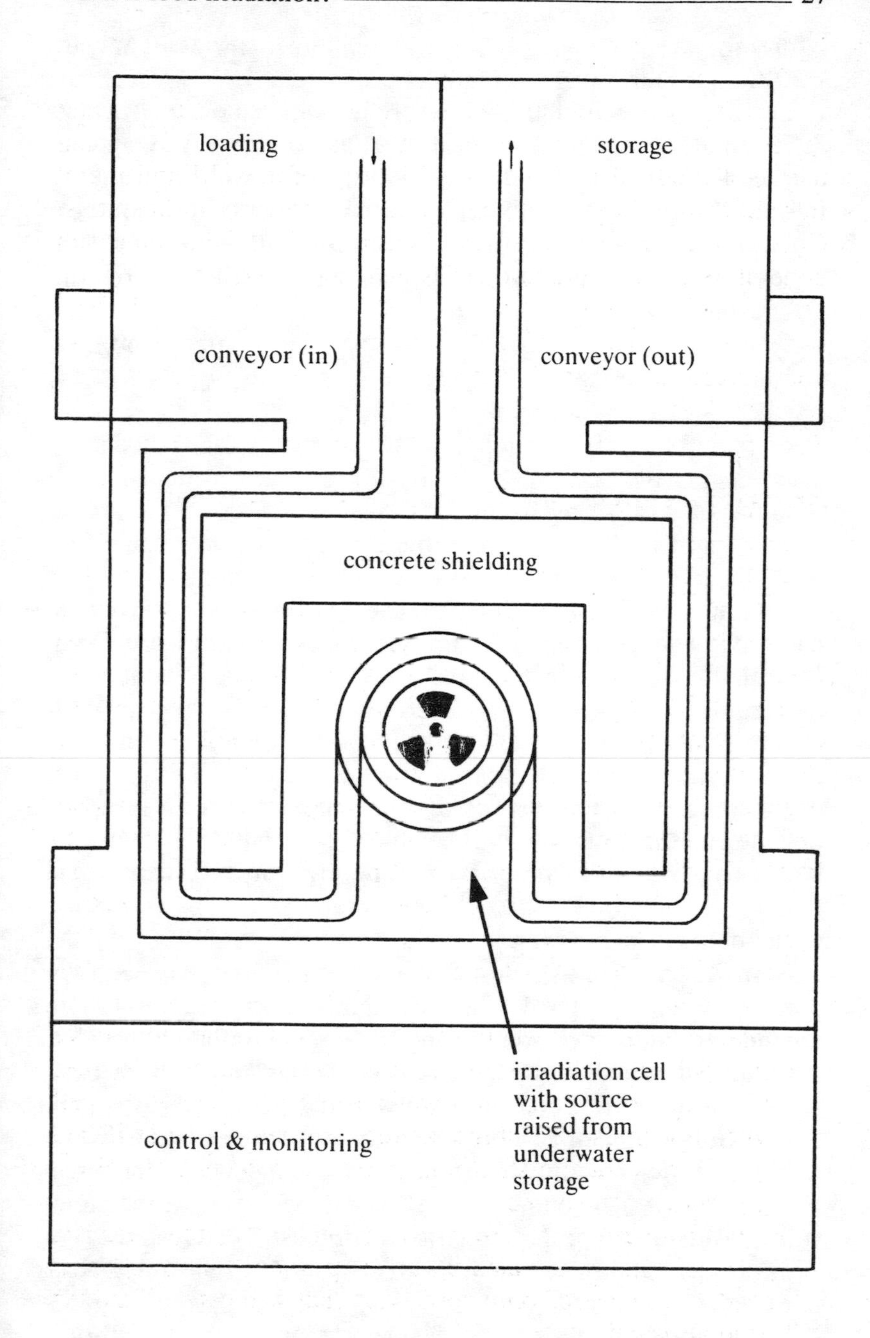

Figure 2: Layout of an irradiation facility

temperatures before being dispatched to the warehouse, food processor or retailing outlets.

- *a control unit* which governs the raising or starting of the irradiation source and the movement of food through the irradiator.
- *a fuel handling unit* for plants using radioactive sources where the sealed rods or strips of cobalt 60 or caesium 137 are received and loaded by remote handling into the irradiator.
- *facilities for monitoring* doses to the irradiated food and keeping records.

The shape of the irradiation cell, the siting of the radioactive sources, and the path of the food through the cell depends on the type of food being irradiated. The time taken to pass food through and its distance from the source will determine the dose received. In the case of a radioactive material source, allowance has to be made for the fact that the source is continually decaying in strength. Some types of irradiator use a central source and pass the food around it on one of a number of circular tracks. Often the arrangement of conveyor and source is more complicated with the food container turned so that it is irradiated from both sides. The size of the package determines the difference between the maximum doses on the outside and the minimum in the centre of the bulk food package being irradiated. In practice, the maximum dose can be as much as two to three times the minimum unless the thickness of the bulk package is small.

Some foods such as fish may pass along a tube between two parallel sources. In the USA, the Atomic Energy Commission considered an irradiator for use on board ships but the pilot plant took up too much space and was moved to a dock instead. If this idea were revived, the plan would be to lightly irradiate fish at sea and then further irradiate when it comes ashore. Other facilities may use a simple conveyor belt past either a gamma-ray or a machine source. The USSR uses a machine source to irradiate imported grain at the port of Odessa. The French have designs for a small-scale field irradiator for potatoes.

There are currently over 130 operating irradiation facilities world-wide. A further 30 are either planned or being constructed. However, not all of these irradiate food and, of those that do, many also handle medical supplies. Appendix A1 lists the facilities known, planned or under construction and indicates those for irradiating food.

The USA has 17 operating food irradiation facilities. Most include relatively small amounts of food (mainly spices) in their total inventory of irradiated products. In 1986, the US Department of Energy (DoE) proposed building six irradiation facilities specifically for food and using caesium 137 as the source. These were to be sited in Hawaii, Washington, California, Iowa, Oklahoma, and Florida. None had been constructed as of 1989. The USSR is currently the largest user of irradiation. A machine source with a capacity of 40,000 tons a year irradiates grain imports at the port of Odessa. In Europe the main facilities irradiating food are the Gammaster plant at Ede, the Netherlands; the IRE plant at Fleurus, Belgium; and in France at the Conservatome plant near Lyons and the Caric plant near Paris. The total capacity of all facilities in the European Community is only 47,500 tons a year — and this if all were solely devoted to irradiating food rather than the medical supplies and other products.

Britain has ten facilities which currently irradiate medical supplies or animal feed. Of these only four, all owned by Isotron plc, will be able to handle commercial food irradiation. One other may be able to do so and privatization of some hospital facilities might open possibilities for others on a small scale.[10 11]

Clearly, irradiation of food is still a technology in its infancy. The total capacity world-wide is small and the actual amount of irradiated food is still trivial compared to the total volume of food traded.

Radiation doses for food irradiation

'Dose' means the amount of energy transferred by irradiation. This is measured in a unit called the Gray. Current proposals for food irradiation would permit doses of up to 10,000 Gray. This is usually expressed as ten kiloGray or 10 kGy. Sometimes an older unit, the rad is used. The 10 kGy maximum is equal to one million rad — or 1 Mrad.

The actual doses given to various foods will vary between about 0.1 kGy and 10 kGy, with occasional uses permitted above this level. These doses are massive. For comparison they are roughly 1–100 million times what a hospital patient would get from a chest X-ray, or the average person would get from natural sources of radiation in the course of a year. A lethal dose for a person exposed would be around 6 Gy — less than one-thousandth of the dose that can be given to food. Clearly this is a dangerous technology where particular care must be taken to prevent exposure of workers or the public.

It is important to distinguish between the dose given to the food and the energy level of the radiation source. At high energy levels it is possible for some elements of the food to be made radioactive. At the energy levels of the two radioactive source materials, or with the restrictions agreed for machine sources, this is not a significant problem. It is proposed that electron beam sources should be limited to energies of 10 million electron volts (10 MeV) and X-ray sources to 5 MeV. The two radioactive sources have lower energies still. Cobalt 60 gives off two gamma rays at 1.17 and 1.33 MeV and caesium 137 gives off a single gamma ray at 0.66 MeV. At these levels, the amount of induced radioactivity will be extremely small and die away very rapidly. It will certainly be undetectable against the background level of radioactivity that already exists in food.[12]

The dose determines the extent of the physical, chemical and biological effects on the food. It also determines the effect on insects and bacteria. Higher doses will increase these effects but will not produce more radioactivity in the food. Higher doses also increase the undesirable effects. Irradiation of particular foods has to be limited to a narrow range of doses — sufficient for the desired effect and not so much that the food product is damaged and made unacceptable.[13]

Uses of irradiation

It is claimed that the effects of irradiation can be used to:

- inhibit the sprouting of some root vegetables
- delay the ripening of some fruits and vegetables
- kill, or render sterile, some insects and parasites

that infest grains, fruits, vegetables and meats
- reduce the levels of spoilage bacteria on some foods, such as fruit, seafood, poultry and meat, enabling them to have a longer shelf-life
- reduce the levels of food-poisoning bacteria on seafoods, poultry, meat and spices.

Table 1 shows the doses recommended by the World Health Organization for various applications. These doses are only

Table 1: Doses for specific food applications[14]

Purpose	Dose (kGy)	Products
Low dose (up to 1 kGy)		
(a) Inhibition of sprouting	0.05-0.15	Potatoes, onions, garlic, ginger-root, etc.
(b) Insect disinfestation and parasite disinfection	0.15-0.50	Cereals and pulses, fresh and dried fruits, dried fish and meat, fresh pork, etc.
(c) Delay of physiological process (e.g. ripening)	0.50-1.0	Fresh fruits and vegetables
Medium dose (1-10 kGy)		
(a) Extension of shelf-life	1.0-3.0	Fresh fish, strawberries, etc.
(b) Elimination of spoilage and pathogenic micro-organisms	1.0-7.0	Fresh and frozen seafood, raw or frozen poultry and meat, etc.
(c) Improving technological properties of food	2.0-7.0	Grapes (increasing juice yield), dehydrated vegetables (reduced cooking time), etc.
High dose (10-50 kGy)*		
(a) Industrial sterilization (in combination with mild heat)	30-50	Meat, poultry, seafood, prepared foods, sterilized hospital diets
(b) Decontamination of certain food additives and ingredients	10-50	Spices, enzyme preparations, natural gum, etc.

*high-dose applications not yet endorsed by the Codex Alimentarius Commission

approximate. The actual dose in commercial applications will depend on a variety of factors. The effects of irradiation can be applied in three basic ways:

- to extend the shelf-life of foods — the time it takes before food stored in warehouses and shops becomes unsaleable — i.e. as a *preservative* to reduce food wastage
- to guarantee that live and fertile insects do not get transported in food to become a pest in areas where they currently do not exist — i.e. as a *pesticide* for quarantine control
- to reduce the levels of bacterial contamination, especially food-poisoning organisms, making the food micro-biologically 'cleaner' and, it is claimed, safer to eat — i.e. as a *disinfectant*.

Improvement of food by irradiation?

A number of other 'improvements' in food quality have been claimed.[15]

- *In bread making*, irradiation is said to improve the baking and cooking quality of wheat, including the ability to add up to 15 per cent soya flour to wheat flour without loss of baking quality. Irradiation also 'improves' the elasticity and volume of dough in bread making. A number of additives are currently used to increase the bulk, and the water and air content, of the standard white loaf. Yeast can be stimulated by irradiation, leading to faster bread making.[16] While this has obvious benefits to the large baking firms, it is a matter of opinion whether it leads to an improvement in bread quality.[17] [18]
- *In brewing, wine making and distilling,* irradiated barley can increase yield during malting by 7 per cent — a fact of interest to the brewing industry. Irradiation can be used to 'age' spirits,[19] and irradiated grapes yield more juice when processed.
- *In food processing*, irradiated sugar solutions can be used as preservatives, possibly replacing other chemicals used for this purpose in processed and prepared foods. The time needed to reconstitute and cook dehydrated vegetables (peas and green beans) is reduced if these are irradiated. Since cooking times for dehydrated foods are already very

> short it is debatable whether this is any real benefit. It is claimed that irradiation enhances the flavour of carrots and suggested that it could be used for 'tenderizing' of meat.[20]

Most of the basic mechanisms of these 'favourable' changes in food quality are not fully understood.[21 22 23]

Other non-food, but food-related, uses include modification of starches for the paper and textile industries and sterilization of gelatin for use in the photofilm industry.

It has also been suggested that contaminated or spoiled food can be sterilized by irradiation and so made safe for human consumption. Despite explicit recommendations from the World Health Organization that irradiation should not be used to make an unsuitable product saleable for human consumption, and that food should always be wholesome before irradiation, there are some who argue that this should be permitted.[24] Indeed, as we will show, there have been numerous scandals involving use of irradiation to conceal unacceptable levels of contamination on unsaleable foods in order to put these back on the market.

On balance, most of the above uses are not necessities. They are either luxuries or techniques which might benefit the manufacturer with no clear benefit for the consumer.

A brief history of food irradiation

In order to understand the present situation, we have found it helpful to review the way irradiation has developed. This history involves many overlapping events. At the end of the book there is a chronology of the major events year by year.

The idea of irradiating food is not new. The treatment was tested on strawberries in Sweden in 1916. The first patents on the idea were taken out in the USA in 1921, and in France in 1930. Little progress was made until 1953, when US President Eisenhower announced the 'Atoms for Peace' programme. Public attention was shifted away from nuclear weapons by promoting nuclear energy and other peaceful uses of nuclear technology. This allowed the academic and industrial infrastructure of nuclear and radiation using technologies to be developed — behind which the weapons programme continued. A decade of intensive research into food irradiation was funded and supervised by the US Department of Defence.[25] [26]

The first commercial use of food irradiation actually occurred in Germany in 1957 for sterilization of spices used in sausage manufacture. This was brought to an abrupt end when the German government banned the process in 1958. The Soviet Union was the first government to permit irradiation, for inhibiting sprouting of potatoes in 1958, and disinfestation of grain in 1959. Canada permitted its use for potatoes in 1960. In 1963 the USA gave clearance for inhibition of potato sprouting and wheat disinfestation. Permission was also given for irradiation of can-packed bacon but this clearance was withdrawn in 1968 after a review of the research found adverse effects and deficiencies in the conduct of the experiments.[27] It was not until 1986 that the USA cleared irradiation for use on fruit and vegetables, for shelf-life extension, and pork, for control of the trichina (trichinella) parasite. These clearances were only up to a maximum dose of 1 kGy. An exception was made for spices which were allowed doses up to 30 kGy for disinfestation/decontamination.[28]

The food industry has shown little interest in developing food irradiation. It was certainly not willing to finance the research necessary to convince governments that the process is safe. Few governments showed any interest in doing this either. Left to itself, irradiation would have remained a vague possibility with little commercial application and with few government clearances. The early data raised more questions than it answered. In the UK, for example, a 1964 report of a scientific working party[29] decided that safety could not be assured and the process was effectively banned in 1967.[30] The only exceptions were for animal feed and, in special cases, for patients needing a sterile diet. In Europe, markedly different approaches to food and other safety issues led to it being banned in West Germany — where there is a tendency for new processes to be illegal unless there are laws permitting them — but permitted in the Netherlands, where things are usually permitted unless laws prohibit them. Until 1990, six of the EC countries permitted irradiation while the remainder either effectively banned it or had no laws governing its use.[31] Such differences between neighbouring countries clearly acted as a barrier to international trade in irradiated food products.

To overcome international inertia, the International Atomic Energy Agency (IAEA) established a Joint Expert Committee on Food Irradiation (JECFI) in 1964 with the UN Food and Agriculture Organization (FAO) and the World Health Organization (WHO). The IAEA has the UN mandate to control and develop peaceful uses for atomic energy. It collaborates with the FAO on uses of radiation in food and agriculture. Food irradiation was an important area for development as it could use waste and by-products of the atomic energy programme. Nineteen countries financed an International Food Irradiation Project (IFIP), based at Karlsruhe in West Germany. This was to co-ordinate research on safety of irradiated food which could be used by the IAEA/FAO/WHO Joint Expert Committee to clarify uncertainties over safety — and there were a lot of them in 1964. The JECFI met in 1969, 1976 and 1980. It produced major reports in 1970,[32] 1977[33] and 1981,[34] giving first conditional clearances for some foods and then a general clearance for all foods irradiated up to a maximum average dose of 10 kGy.[35] The IFIP was wound up in 1982.

In spite of the assurances of the Joint Expert Committee, governments continued to move cautiously, if at all, in granting clearances for irradiation. Of some 160 countries world-wide, only around 35 countries have permitted irradiation for public consumption and most of these only for a limited selection from a total of about 30 different foodstuffs. Table 2 provides a summary of these clearances. A more detailed list of the permits by each country and the date they were given can be found in Appendix A2.[36] [37]

Several significant nations, such as West Germany, Sweden, Australia and New Zealand, do not permit irradiation. In others the permission is still extremely limited, notably Japan (only for potatoes) and Denmark, Norway and Finland (for spices). Many countries which permit its use only for a limited range of foodstuffs do not have irradiation facilities and, even among those which do, the total volume of irradiated food is trivial (see Appendix A1).

It was only in 1983 that any public opposition to irradiation began to surface, initially in Canada and the USA.[38] [39] In the UK there was little awareness of the issue until 1985, when a working group on food irradiation was established at the London Food Commission. We presented a series of unresolved

Table 2: Irradiation clearances – breakdown by country and major food category

	Potatoes	Onions & shallots	Garlic	Other fresh vegetables	Semi-dried and dried vegetable products	Prepared vegetables, pastes and fillings	Strawberries	Exotic fruits (incl. mangoes, papayas lychees, avocados)	Dates/dried fruit	Ginger	Other fresh fruit (incl. tomatoes)	Frozen fruit juice	Herbs & spices (incl. black and white pepper, tumeric, paprika)	Cocoa beans	Chocolate	Edible nuts and seeds
Yugoslavia	●	●	●		●				●							
Uruguay	●															
USSR	●	●							●		●					
USA	●	●											●			●
Thailand	●	●	●				●	●	●		●		●	●		
Spain	●	●														
S. Africa	●	●	●	●	●	●	●	●	●	●	●	●	●			●
Poland	●	●														
Philippines	●	●	●													
Norway													●			
Netherlands	●	●		●	●	●	●						●	●		
N. Zealand																
Mexico		●										●	●		●	●
Korea	●	●	●		●											●
Japan	●															
Italy	●	●	●													
Israel	●	●	●	●	●		●				●		●	●		●
Indonesia	●	●	●										●			
India	●	●											●			
Hungary	●	●		●			●				●		●			
GDR		●											●			
France	●	●	●		●								●			
Finland													●			
Denmark													●			
Czech	●	●		●												
Cuba	●	●	●													
China	●	●	●	●												●
Chile	●	●					●	●	●				●	●		
Canada	●	●											●			
Bulgaria	●	●	●						●		●					
Brazil	●	●		●			●	●					●			
Belgium	●	●	●		●		●		●		●		●			
Bangladesh	●	●						●					●			
Argentina	●	●	●				●									

Table 2 continued

	Argentina	Bangladesh	Belgium	Brazil	Bulgaria	Canada	Chile	China	Cuba	Czech	Denmark	Finland	France	GDR	Hungary	India	Indonesia	Israel	Italy	Japan	Korea	Mexico	N. Zealand	Netherlands	Norway	Philippines	Poland	S. Africa	Spain	Thailand	USA	USSR	Uruguay	Yugoslavia
Rye Bread																								●										
Malt																								●										
Wheat and wheat products		●		●	●	●	●	●																						●	●			
Other cereals and cereal products				●	●	●		●					●				●	●														●		●
Rice		●		●	●		●	●																●						●				
Pulses		●					●											●												●				●
Teas, tea extracts																												●			●			●
Poultry		●		●		●	●						●					●						●				●		●	●			●
Frogs' legs		●														●								●										
Sausages								●																						●				
Other meats and meat products																								●						●	●	●		
Fish (inc. fresh and frozen shrimps)		●		●		●	●									●								●						●				
Other products*				●									●	●										●				●			●			●

* include gum arabic, cheese powder, yeast powder, egg powder, soya pickle products, dry blood protein, enzyme solutions and dry or dehydrated enzyme preparations and powdered batter mix.

Source: based on information from IAEA/WHO1988

Note: Within the broad food groups given above, national clearances have generally been given for very specific products which were too numerous to list.

questions[40] to the UK government's Advisory Committee on Irradiated and Novel Foods (ACINF)[41] in 1985. Two days before this committee reported in 1986, we helped uncover a major scandal involving the use of irradiation to cover up contamination of unsaleable foods. At the same time, opposition MPs tabled a House of Commons motion questioning possible conflict of interest in the activities of one of the technical advisors to the UK Advisory Committee (see Appendix B1). Later that year, public concern over all things nuclear was heightened by the Chernobyl nuclear accident in the USSR. These events, combined with growing public awareness as a result of activities of our working group, led to an unprecedented critical public response to the idea of irradiation. The UK Advisory Committee received over 6,000 comments, most of them critical, from a wide range of organizations and individuals. The public concern has continued to grow rather than be calmed by government assurances. Scandal has followed scandal and effectively set back plans for the introduction of irradiation in Britain to 1990.

In Europe, the European Commission intends to remove all barriers to trade between EEC countries by 1992.[42] It has chosen to interpret different national controls on irradiation as an obstacle to trade and proposed harmonizing legislation to permit its use for a wide range of foodstuffs. The European Parliament has so far declined to agree with the Commission. The Parliament voted against general approval in 1987 'on precautionary grounds'[43] and in 1989 it voted for a general ban on irradiation with a special exception for spices only.

Since 1983, the promotion of food irradiation has been led by IAEA/FAO International Consultative Group on Food Irradiation (ICGFI). By 1988 this promotion work was being supported by governments of 26 countries (see Appendix C3). Late in 1986, this Consultative Group convened a task force meeting of public relations and marketing experts to advise on how best to promote acceptance of irradiation.[44] We will analyse the marketing strategy which emerged from this meeting in more detail in Chapter 6. The key elements involve convincing the world's public that irradiation is vitally necessary to deal with two critical food problems — food poisoning and world hunger. The strategy calls for concerted efforts to promote acceptance of irradiation by governments, industry, non-governmental organizations and consumers.

Some elements of this strategy are already in place, others are being developed. Some elements have failed. Environment, consumer, women's, public health, trade union and some food industry organizations have rejected the arguments put forward for irradiation. In 1987 the World Congress of Consumers Organizations unanimously called for a moratorium on any further use or development of irradiation until a number of key issues are resolved (see Appendix C2). In 1988 an international UN conference failed to get agreement from participating countries on promoting acceptance of irradiated food. Concerns of consumer and environmental organizations were taken up by a number of government delegations. A proposal to endorse the statement that there were 'no unresolved safety issues' had to be withdrawn. Several major countries dissociated themselves from the final resolution.[45]

The questions of global concern

This then is the global picture. Considerable effort is being made by some international agencies to promote food irradiation as safe, beneficial and necessary. There is resistance on the part of some countries and reluctance to act on the part of others. There is a general lack of interest in food irradiation on the part of the food industry. There is considerable criticism and organized resistance from many consumer, environmental, public health and trade union organizations. There is growing concern among the general public as emerging scandals contradict the image that the promoters of irradiation would like to foster. The stage is set for a struggle to win the hearts and minds, wallets and purses of the public, not just in the UK but world-wide.

The authors of this book are not neutral observers. We have been actively involved in this struggle since 1985. The issues we raised in 1985 have been taken up world-wide. How is it that governments can be so divided on the advice and recommendations of international experts? On what basis do the consumer, public health and environment movements challenge this advice? Behind all the arguments and assurances, what are the facts? What is wrong with irradiation?

We believe that there are three main questions that need to be answered before irradiation is permitted.

- Is irradiated food safe and wholesome?
- Is the technology controllable in the real world of the international food trade?
- Is it really needed — and, if not, who wants it and why?

If there is any doubt about these questions, there is a crucial fourth question.

- How can the general public ensure that further development of irradiation is halted until these doubts are resolved?

It is these questions we will address in the chapters which follow.

3. Is it safe to eat?

According to its promoters, the process of irradiation has been more thoroughly tested than many other food processing techniques and there are no unresolved safety questions. Provided the process is properly controlled, irradiated food is both safe and wholesome. As the United Nations Expert Committee put it,

> The irradiation of food up to an overall average dose of 10 kGy presents no toxicological hazard and introduces no special nutritional or microbiological problems.[1]

This assessment is cited as the final word on the subject by many governments. The UK government's Advisory Committee on Irradiated and Novel Foods,[2] the European Commission's Scientific Committee for Food[3] and a Danish food agency working group[4] also concluded that, within limits on the applied dose and the energy of the radiation source, there were no special safety or wholesomeness problems.

The US Food and Drug Administration (FDA) has granted only limited approval for irradiation up to 1 kGy. In 1986 it held that approval for irradiation above this level might require further safety testing.[5] This is because, in the USA, irradiation is governed by the 1958 Food Drug and Cosmetics Act which classifies irradiation as an 'additive' and requires users of irradiation to petition the FDA for clearances for specific foods.[6] The FDA requires users to show by specific evidence that the food will be safe. Wheat, pork, fruit and vegetables were cleared up to 1 kGy but spices were allowed up to 30 kGy. The FDA insists that the stricter 1 kGy limit should not be interpreted as implying that it believes there are safety problems above this level.[7]

Who then are we to challenge these international experts? Unfortunately, virtually every hazard, be it asbestos, thalidomide, agent orange, industrial chemicals, food additives and even radiation itself, has been declared safe at some time. There is often a 20-30 year gap between available evidence of harm and the official recognition of the risk to the public.[8] The experts are not infallible. Traditionally science is slow to recognize potential dangers. In weighing the balance of evidence for and against there being a hazard,

- experts frequently disagree over the validity of the evidence or how it should be interpreted
- evidence can be selected, manipulated or misrepresented to support a particular view
- the scales often appear to be weighted in favour of the hazardous agent rather than the health of the public.

It is the duty of independent consumer and public health watchdog organizations to take a critical view of the evidence — to seek the facts and ask questions about the processes used in making judgements about safety. We would be irresponsible if we accepted without question the opinions of experts, however highly placed. Society as a whole needs to learn from past mistakes and demand a much more stringent system for testing the safety of new products and processes. Having examined the issue in some detail, we have to say that the public would be most unwise if it were to accept the expert advice on the safety of irradiated food uncritically. Assurances of safety have been offered: evidence of safety has yet to be presented in an acceptable form.

Scientific reports indicate a wide range of adverse effects from feeding irradiated food to animals. The way this evidence is dealt with by the various expert committees is extremely disturbing. Far from providing reassurance, our investigations suggest that the whole issue needs to be opened for public scrutiny and the experts called upon to justify some of their biased, misleading, illogical and inaccurate statements. This needs to be done now — before irradiation is given widespread clearance.

Irradiation is still a technology in its infancy. There is every reason to demand a stricter standard of safety testing and evaluation of the scientific evidence than has been adopted in the past. Those who offer assurances about safety of irradiated food have yet to realize that the rules for consumer acceptance of their opinions have changed. In the 1990s they are very different from those of the 1950s, 60s and 70s when the research on which they base their opinions was undertaken.

We are not alone in this view. The British Medical Association's Board of Science said of the UK government's Advisory Committee report:

> The Board believes that the current advice, as set out in the Report on the Safety and Wholesomeness of Irradiated Foods, may not sufficiently take account of, still less exclude, possible long-term medical effects on the population, given that irradiated products have been available for a relatively short time. More scientific data is required . . . and the Board believes that a full-scale study should be undertaken in collaboration with the medical associations of those countries where the process is already in use. Such a study is necessary before the process can be confidently accepted in this country.[9]

The UK Advisory Committee itself suggested a need to monitor patterns of consumption of irradiated foods and their nutrient content to detect any unforeseen nutritional consequences, and to continue the review of new safety data.[10] It also acknowledged areas where there was insufficient scientific information, such as the effect of irradiation on food additives, contaminants, pesticide residues and packaging materials.[11] Unfortunately, it did not see fit to commission this research before granting approval to the process. The UN Expert Committee reports also highlight areas where further work is recommended.[12]

What is safe?

'Safe', as far as various expert committees are concerned, is the absence of any special or significant effects from eating irradiated food. These effects can be evaluated scientifically in four different disciplines:

- *radiation physics and chemistry* — the radiological and chemical effects of irradiation on the food, with particular emphasis on the question of induced radioactivity and the nature of the 'radiolytic' chemicals created
- *toxicology* — the evaluation in animals of the effects of eating these chemical products of irradiation
- *nutrition* — the evaluation of the possible effects on human health of any damage done to essential nutrients in the food
- *microbiology* — the effects on health of changes in the balance of bacteria, yeasts, moulds and viruses.

We will examine each of these areas in the chapters which follow. We will also show how analysing safety according to these separate disciplines has, in some ways, served to confuse rather than clarify the question 'is it safe?' Irradiated food and its effects on health need to be viewed as a whole — and viewed not just in a laboratory setting, but in the real world. We will also analyse the process and the arguments which the various expert committees have used to resolve the problem of conflicting evidence and show why this gives us cause for concern.

Is it radioactive?

There is one area where there is general agreement about irradiated food. That is, provided the limitations on the energy of the radioactive sources are observed, there should be no significant risk of the food being made radioactive. Bombarding food with high-energy ionizing radiation can cause some elements within the food to become radioactive. At the energy levels proposed for food irradiation, the amount of radioactivity created will be very small and die away rapidly.[13] [14] The UK Advisory Committee suggests that, as a precaution, food should be stored for 24 hours after irradiation before it is eaten.[15] We agree that this is sufficient. Any residual radiation would be undetectable against the 'background' level of radioactivity in food.[16] Just as people are not made radioactive by having a chest X-ray, so irradiated food is not radioactive. Neither the food nor the people who eat it are likely to glow in the dark. The only way that food could become radioactive is through damage to the radioactive source in the irradiation cell which resulted in contamination of the food being irradiated. Obviously, great care will be taken to prevent this kind of accident.

Toxicology: the chemical safety of irradiated food

There, however, the agreement ends. Irradiation produces chemical changes in food. The result is a wide variety of what are called radiolytic products.[17] There is considerable controversy over how these might affect people who eat irradiated food and future generations. Some of the changes resemble those which occur in other forms of food processing.[18] Some are unique to irradiation.[19] [20] [21] Some of these chemical changes are known to be harmful, both mutagenic (altering genetic structures) and carcinogenic (causing cancers).

In particular, irradiation produces what are known as 'free radicals' — highly reactive parts of the original chemical constituents of the food. Free radicals are believed to be common cancer promoters[22] — that is, they promote the second stage developments that turn initially damaged living cells into malignant (i.e. cancerous) ones. They are implicated in accelerating the ageing process and lowering resistance to disease.[23] The exact mechanism is not fully understood but may be related to the way that some vitamins are used up in terminating the free-radical chain reactions. Most free radicals react rapidly with other chemical constituents in the food and produce stable radiolytic products. Some of these may be harmful. In some foods, particularly food with high concentrations of polyunsaturated fats, the UK Advisory Committee acknowledges,

> the free radicals formed by irradiation initiate self-propagating chain reactions which continue for some weeks after the irradiation is finished and may lead to higher levels of some radiolytic products being present in certain irradiated foods.[24]

The processes are known as peroxidation and epoxidation and lead to what is commonly referred to as 'rancidity'. The oxidation processes are normally arrested by 'antioxidants' in the food. These can be either chemical additives or natural vitamins such as vitamin E, vitamin C, vitamin B2 (riboflavin) and vitamin B3 (niacin). Sulphur-containing compounds such as glutathione, and trace minerals such as selenium, also aid the antioxidant process. In addition, vitamin A and its precursor beta-carotene[25] protect against carcinogens in the food. Each of these vitamins, as we will show in the next chapter, is damaged by irradiation. Irradiation thus creates potentially harmful products and, at the same time, reduces the natural and essential components in food that assist the body in fighting disease.

To the promoters of irradiation and the various expert committees that have reviewed the issue, these are not significant concerns. The UK Advisory Committee concludes that,

> there are no toxicologically significant qualitative differences between the radiolytic products in irradiated foods and the products in conventionally processed foods (i.e. those processed by cooking, smoking, and treating with chemical preservatives).[26]

Testing chemical safety

Many people question the safety of some food additives and the residues in food of fertilizers and pesticides used in agriculture. The established procedures for testing the safety of these chemicals involve feeding animals large quantities of the suspect chemical and observing any health effects. Since irradiation produces chemical changes in food, should these changes be regarded as additives? If so, how should they be tested?

The view of the national and international expert committees is that irradiated food should not be tested in this way. To do so would require all radiolytic chemicals to be identified, isolated and fed separately in large quantities to test animals. Because of the complexity of the chemical reactions, it is difficult to do this. The UK Advisory Committee states,

> It is not possible to exclude the possibility that unique and potentially toxic substances might be formed in some irradiated foods. We consider that this situation is *unavoidable* since the identity and concentration of all the radiolytic products which could be formed in any of the food which might conceivably be irradiated is not yet known. . . . The inevitable limitations of analytical techniques mean that our knowledge of radiolytic products is incomplete.[27]

Apart from being time-consuming and expensive, the tests would almost certainly show adverse effects since some of the radiolytic products are known to be harmful. So, incidentally, would some of the products of cooking, especially barbecued food. The concentration of some of the radiolytic products can be reduced after storage.[28]

Testing the safety of irradiated food has relied on simply feeding irradiated food to experimental animals. However, in normal toxicological experiments there is a safety margin. One feeds chemicals in doses 100 times greater than they would normally be eaten. With irradiated food one could use higher doses to produce 100 times the level of radiolytic chemicals. However, if one does this the food is unpalatable. The animals are simply unable to eat it. Even relatively small increases in radiation dose produce quite marked changes in the flavour and texture of the food. Fats smell and taste rancid, fruits and vegetables turn to mush. Commercial irradiation of food for human consumption will use doses such that these effects are not noticeable, or can be concealed by added flavourings but, in toxicological science, these act as a serious barrier to safety testing. The UK Advisory Committee admits that if the normal 100-fold safety margin were required then irradiated food could only be evaluated as safe at levels where it comprised less than one quarter of one per cent (0.25 per cent) of the human diet.[29]

Testing without a safety margin may miss underlying or long-term safety hazards. A review for the Hungarian Academy of Sciences in 1979 points out that even very small quantities of some of the radiolytic chemicals could be hazardous.[30] With chemicals that can cause cancer or genetic defects, it is safest to assume that there is no safe level of exposure. Any exposure can cause the initial damage that develops into a cancer years later. Damage to the genetic blueprint may cause miscarriage or defects in future generations. The fact that only a small amount of a toxic chemical is present does not eliminate the risk;[31] it merely reduces it, as we are discovering with some of the chemicals that have been approved as 'safe' by the conventional methods.

Some studies have attempted to build in a safety margin by irradiating with doses in the 10-50 kGy range. The results here are less than reassuring. It is unclear whether the adverse effects are due solely to the unpalatability of the food (and resulting malnutrition) or to other underlying toxicological problems. The expert committees have chosen to dismiss the evidence of harm from these studies, saying they have no significance for safety of irradiation at doses below 10 kGy. At the same time, many promoters of irradiation insist that there is no established evidence of harm above this level either.[32] More recently, some radiolytic products have been isolated and tested in the usual manner for additives.[33] The results are claimed to be reassuring. Even though some adverse effects were found, it is claimed that these are not likely to occur under practical conditions.[34]

Adverse effects

Adverse effects reported from the animal feeding experiments include:

- lower growth rates
- breakdown of muscle tissue
- increased incidence of tumours
- heart lesions and problems with blood clotting
- kidney damage
- lowered immune response
- chromosome defects
- reduced numbers of offspring
- low birth-weights in offspring
- mutations in offspring
- miscarriages
- damage to testicles and sperm cells.

A 1979 a review of food irradiation literature by J. Barna for the Hungarian Academy of Sciences[35] classified the result of the published studies as neutral, adverse, or beneficial. Barna found 7,191 neutral effects, 1,414 adverse effects and 185 beneficial effects For example:

- bacon: 86 neutral study results, 31 adverse study results, and no beneficial results.
- soyabeans: 26 neutral study results, 60 adverse study results, and no beneficial results.
- sucrose: 38 neutral, 39 adverse and 1 beneficial.
- corn oil: 5 neutral,13 adverse and no beneficial results.

The report concluded that Barna's evaluation of the data on the wholesomeness of irradiated foods leads to the conclusion that, at present, the effects are inconsistent, ambiguous and unreproducible. These effects cannot be traced back to a given food or group of foods or level of radiation dose.

In addition Barna noted that feeding irradiated foods showed a *tendency* to produce adverse effects — which was discounted because they were not statistically significant — but that,

> In some cases the authors *did not attach biological or toxicological significance even to statistically significant changes*. They demanded *further* investigations to *confirm* adverse findings registered in their feeding studies.[36] [emphasis in original]

Reading the original of Barna's report, one is left with a worrying feeling that the scales of uncertainty are weighted against the studies suggesting possible harm. In toxicology, 'significance' involves mathematical analysis of the results to see if the change in the number of cases of some health indicator (e.g. tumours) was likely to have been caused by the chemical or just by chance. An adverse change is significant only when the probability of it being due to chance is less than one in twenty. Many examples of changes where the risk of being wrong is only one in ten are simply called 'not significant'. Being right nine times out of ten is not good enough. If the test were applied in reverse — that is, all studies be required to demonstrate positive proof of safety with a probability of not being wrong more than once in twenty cases — then irradiated food would probably fail. Clearly some balance has to be struck, but where — as Barna indicates — there is considerable uncertainty, which should be given the benefit of the doubt: the public health or the process of irradiation? The root problem lies in testing without the normal safety margin. This method is clearly not sensitive enough to pick up the underlying problems.

There are studies that did not find some of the adverse effects noted above.[37] The World Health Organization claims that none of the studies carried out under the auspices of the International Food Irradiation Project showed any indication that irradiated foods contained any carcinogens or other toxic substances.[38] The emphatic nature of this statement is a gross oversimplification of the scientific evidence provided by the

IFIP. In any case, direct comparisons with studies that did find problems are difficult as the experiments were often not conducted under identical conditions. In addition, many of the studies, on both sides of the argument, were badly designed, poorly conducted or inadequately reported.

Fraudulent research

Some studies are suspected of being fraudulent. Early research done by the US Army from 1953 was used to obtain the clearance for can-packed bacon in 1963. This clearance was subsequently withdrawn in 1986 when the US Food and Drug Administration found the research was flawed. Significant adverse effects, such as fewer surviving offspring, were produced in animals fed irradiated food. The Food and Drug Administration also found major deficiencies in the conduct of some experiments.[39]

Subsequent research done by Industrial Biotest Limited was also flawed. Three officials of the company were convicted in the US courts in 1983 for doing fraudulent research for government and industry. Problems such as failure to conduct routine analyses, premature death of thousands of rodents from unsanitary laboratory conditions, faulty record keeping, and suppression of unfavourable findings were uncovered. Prior to the convictions, the Army had declared the beef and pork feeding studies being conducted by the company in default for similar contract violations. The Army discovered:

> missing records, unallowable departures from testing protocol, poor quality work, and incomplete disclosure of information on the progress of the studies.[40]

The government lost around $4 million and six years worth of animal feeding study data on food irradiation.

In 1986, the US Food and Drug Administration evaluated over 400 studies using objective criteria for judging whether they were adequately or inadequately conducted. The result was that all but five studies had to be discarded. Even these suggested there might still be problems. The director of the Food Safety Laboratory at the US Department of Agriculture's Research Service said of the extensive animal feeding trials conducted by Raltech Scientific Services,

> Two of the studies . . . had some possible adverse findings which will require careful consideration before the process can be declared safe.[41]

In studies of mice fed diets containing irradiated chicken from birth to death, survival of both sexes was significantly reduced and the animals eating irradiated chicken also had the highest incidence of tumours. The US National Toxicology Program assembled a new panel to review the results. This panel declared that the tumours were not cancerous but failed to explain the increased number of deaths among the animals eating irradiated chicken.

Although not highlighted in the US Department of Agriculture review, another of the studies to explore the long-term effects of feeding irradiated chicken had to be terminated prematurely due to excessive mortality among offspring in all diet groups. The study, intended to last two years, was cancelled after only nine months. Instead of repeating the study to determine the true long-term reproductive effects, the researchers declared the process safe on their limited nine months' data.[42]

Fruit flies *(Drosophila melanogaster)*, which are commonly used to test for mutations, were also fed irradiated and non-irradiated diets containing chicken. For comparison, another group was fed a diet containing a chemical known to cause mutations. In scientific terms this is known as a positive control. The Department of Agriculture reported:

> ...an unexplained significant reduction in the production of offspring in cultures of *D. melanogaster* reared on gamma-irradiated chicken. This response was dose-related and was not overcome by the addition of vitamin supplements.[43]

In fact, the fruit flies fed the diet containing irradiated chicken had seven times fewer offspring than those fed unirradiated (cooked) chicken. Food irradiation promoters have declared the results irrelevant, since fruit flies don't normally eat chicken. The data shows that fruit flies consuming no chicken had far more offspring but, comparing the groups eating the chicken diets, the groups fed irradiated diets had fewer offspring than even the group fed the mutagenic chemical.

Initially, promoters of irradiation had argued that it should be treated as a process and not as an additive. This would have reduced or removed the safety testing requirements altogether. This view was rejected. However, the clearance given by the US Food and Drug Administration in 1986 was, in fact, not based on any of the animal feeding studies. The FDA argued that it should be based on a theoretical estimate of the number and amount of unique radiolytic chemicals created. The FDA claimed that, up to 1 kGy, the level of these products was so low as to be acceptable.[44]

Changing of the rules for safety testing does not inspire confidence. The safety testing of irradiated food has been poorly organized. There is conflicting evidence and many safety questions remain unresolved. Nevertheless, the national and international committees have concluded that irradiated food is safe. Is it unreasonable to ask how they have made sense of such a mass of inconclusive and conflicting data?

Opinion or fact?

It is one of the fundamental principles of science that an opinion be backed up by references to the scientific studies which support it. This enables independent researchers to review the data and the methods used in the experiments. It is therefore disturbing to find that the expert committees, on whose reports various governments rely, appear either unable or unwilling to provide proper references to the scientific literature to support their opinions.

The UK Advisory Committee report[45] is unreferenced. All it provides is a bibliography — a list of some relevant publications — but no indication of which studies contain evidence to support the statements in the report. We pointed this out in 1986 and requested detailed references to the studies actually reviewed by the Committee as they related to each of its opinions on safety. These were refused. A similar request from leading Parliamentary Opposition spokespeople met with a similar response.[46] Why? We cannot conceive of how the Committee could have undertaken its four-year investigation without some systematic plan for identifying and reviewing the scientific evidence which provides evidence for and against potential safety problems. This would have been required of a university undergraduate student. It certainly should be expected of a scientific committee advising the government on an issue of public concern.

Several times, in debate on public platforms around Britain, we have asked members of the Advisory Committee to explain how they undertook their investigation. Did they, for example, identify each of the studies which had found adverse effects, then identify comparable studies which had not found the effect and make some evaluation of the strengths and weaknesses of the two sets of studies? If the review was not done this way, then how? If this had been done and the results clearly presented, then it might be possible to conclude that, on balance, irradiated food could be considered safe. But if this had been done there should be detailed papers setting out the evaluation and there should be no reason why these could not be made public. We have yet to receive an answer to this simple question. The silence of the members of the committee is disturbing. They cannot claim that detailed referencing is not the practice with such committee reports. A 1964 report to the UK government on irradiated food was referenced in detail. This report, which was the basis for the ban on irradiation imposed in the UK in 1967, found that there were many unanswered questions needing to be addressed.[47] Unfortunately these detailed questions do not appear to have been followed up by the later committee.

The UK government is not alone in failing to meet this fundamental scientific principle. The Reports of the United Nations agencies' Joint Expert Committee[48] are also unreferenced, as is the report of the European Commission's Scientific Committee for Food.[49] The best that can be said of these reports is that they offer the opinion that there are no 'special' or 'significant' safety problems associated with consumption of irradiated food if it is properly controlled, but that they have not, so far, seen fit to back up the statements on safety with references to the scientific evidence, nor to recommend adequate systems of control.

There is another safeguard that science has developed which it appears is also being violated. Research papers are submitted to recognized scientific journals where they are subjected to peer review before publication — that is, they are reviewed by independent experts in the field.

The UN agencies' Expert Committee relies heavily on data

provided through the International Food Irradiation Project based at Karlsruhe in West Germany. This was funded by the governments of 24 countries and operated from 1970 to 1982. It was expected to gather together the best available evidence world-wide and to commission new research as needed. Feeding studies were conducted with irradiated wheat flour, potatoes, rice, fish, mangoes, spices, dates and cocoa powder. In all, some 500 studies were undertaken.[50] Much of this and other work reviewed by the UN Expert Committee has not been published. A Canadian survey of the data base cited by the Committee indicates that over 60 per cent of the research has not been published in the open literature. Most appears in unpublished reports, personal communications, or the in-house publications of the UN agencies themselves.[51] Those publications that do exist have not been collected in a central data bank and it is not possible to make a systematic search of the information they contain.[52] Scientific monographs exist which provide more detail than the Expert Committee reports. Peter Elias, former director of the project, conceded that even these did not fully meet our concern over the need for a comprehensive referenced report covering all aspects of the safety of irradiated food.[53]

Bias, illogicality, misrepresentation and inaccuracy

It might be thought that the UN agencies represent the best of the world's independent experts. That being so, it is disturbing to find serious inaccuracies in the publications of these agencies. We will outline some of the worst examples of inaccuracy and misrepresentation in the sections which follow.

An Australian Parliamentary Committee spent three years, from 1985 to 1988, investigating all aspects of food irradiation. It had access to expert advice and evaluation of all the evidence available on safety. Its report, delivered to the Australian government in November 1988, recommends that the World Health Organization be asked to re-open the investigation of the safety of irradiated food and to produce a properly referenced scientific report.[54] This recommendation has been endorsed by the International Organization of Consumers' Unions, the UN-accredited body representing the world consumer movement.

The UN Joint Expert Committee is currently reviewing the safety data on irradiation of food above 10 kGy. We believe that this provides an excellent opportunity to produce the properly referenced, critical evaluation of the lower dose data that is needed. We cannot understand why, if the case for safety is so strong, there is such reluctance to present it properly. If it cannot be presented in this way then there should be an immediate halt to any moves to promote irradiation world-wide.

Perhaps part of the reason why there is such reluctance on the part of the national and international agencies to re-open the investigation is that it would inevitably expose some of the errors in the earlier opinions. In exploring some of the conflicting claims on the safety of irradiated food we have uncovered what appears to us to be systematic bias in the way that evidence is treated. Evidence suggesting irradiated food may be unsafe is treated more critically than evidence that it is safe. We have also uncovered statements that defy normal logic used to dismiss any possibility of harm. We have found opinion misrepresented as fact. We have found the results of some studies misrepresented. We have also uncovered what appears to be either gross scientific incompetence or deliberate scientific fraud.

The National Institute of Nutrition studies

It is sometimes illuminating to examine a specific area of concern in some detail as it can throw light on the way that the whole process of study, review and promotion of biased opinion operates. One such area has been the way the various expert committees have dealt with the results of studies undertaken in the mid-1970s at the National Institute of Nutrition (NIN) in Hyderabad, India.

The NIN studies took as their starting point the approval given by the UN Joint Expert Committee for irradiation of wheat, with the proviso that it should be stored for three months between irradiation and consumption.[55] The experimenters reasoned that, in a country like India, this restriction might present a problem in times of food shortage.

They therefore set out to investigate if there was any problem with feeding *freshly* irradiated wheat.

The test diets were selected to be adequate but nutritionally marginal. Each contained 70 per cent wheat from one of the following categories:

- freshly irradiated
- irradiated but stored for three months
- unirradiated.

Test animals included rats, mice, and monkeys. There was also a small trial involving a group of malnourished children. Much of the public attention has been focused on this latter group for obvious reasons but the main study involved the animal trials. The study of the children merely confirmed one of the adverse findings — that freshly irradiated wheat appeared to cause chromosome damage.

The studies were carefully designed in accordance with standard cytogenetic methodology. In particular:

- comparable animals were used in each test group
- the studies were undertaken 'blind' in that the experimenters did not know which animal or group the microscope slide they examined came from, thereby eliminating experimenter bias
- they tested whether the effects were the result of malnutrition or directly related to the differences in the diet.

The results were disturbing. The animals fed the freshly irradiated wheat were significantly less healthy than those given either the unirradiated or the irradiated but stored wheat diets. The adverse effects noted were:

- lethal mutations occurring in the offspring of rats and mice[56 57]
- damage to the sperm cells of male rats [58]
- increased numbers of foetal deaths in the uterus of mice — the mouse equivalent of miscarriages[59]
- lowered immune response in rats[60]
- polyploidy, a chromosome defect in which multiples of the normal number of chromosomes was found in the bone

> marrow of rats[61] and mice,[62] in the blood of monkeys[63] and the children.[64]

In all cases the changes were found in the animals fed freshly irradiated wheat and the effects were statistically significant when compared to the groups fed unirradiated or stored irradiated wheat. They were also consistent across several species of mammals, including the children, showing that the problem was associated with feeding wheat shortly after irradiation but not if it was stored for three months. The increase in chromosome aberrations was found with the freshly irradiated wheat whether the animals were well-fed or malnourished. The number of polyploid cells increased as the feeding trials continued and was significantly elevated after six weeks. In the case of the children the study was immediately terminated and the children's health monitored. Several months later the number of polyploid cells returned to normal. The animals were fed for 12 weeks. A similar rise in abnormal cells during feeding, followed by a fall after the irradiated food was withdrawn from the diet, was noted.

Other institutes have also investigated this issue. In India, the Bhabha Atomic Research Centre (BARC), which was petitioning the Indian government for permission to use irradiation, and the Osmania University (OU) at Hyderabad, India, published findings of studies which did not observe some of these effects.[65][66] Subsequently, the International Food Irradiation Project commissioned research which also claimed not to have found the polyploidy effect.[67] In a UK study, two out of four groups of mice fed different but irradiated diets showed increases in lethal mutations.[68] The experimenters reported,

> There was some evidence that irradiated (diet) has a weak mutagenic activity . . . which appears to be lessened on storage.

A West German study on Chinese hamsters found a similar rise and fall in polyploidy with the animals fed freshly irradiated wheat.[69] In this case the irradiation dose used was much higher but an increase was observed within one day of feeding (after a day of starvation). No increase was noted with wheat stored for six weeks.

Promoters of food irradiation have gone to extraordinary lengths to discredit the National Institute of Nutrition studies. Much of the attention has focused on polyploidy. The long-term significance of this chromosome defect is not clear but it cannot be dismissed. Normal human cells have a set of 46 chromosomes which contain the blueprint for the cells' functions. Abnormal or extra chromosomes can cause a cell to malfunction. With polyploidy there may be two or three or more sets of chromosomes. This condition is rare in human cells. It can result from failure to separate at cell division and is frequently seen in cancer cells, though this is only one example of the bizarre cell forms present in tumours. Polyploid lymphocytes have been found in increasing numbers with age and during viral infections. As yet, there is no evidence of a direct correlation with cancer incidence.[70] Scientists do not fully understand what mechanisms come into play in the formation of cancer but it is widely believed that damage at the cell level and, in particular, damage to the chromosomes initiates the process. Exposure to a secondary agent may be needed to promote the cancer.[71] As we noted earlier, irradiation creates free radicals which are cancer 'promoters'. The discovery of chromosome damage in the bone marrow and blood of five animal species and in children fed irradiated diets is clearly cause for concern, though not in itself proof that eating irradiated foods causes such effects. It does require careful evaluation of all the scientific evidence. With the other adverse health effects noted in the NIN studies, there is no such uncertainty over the implications. Lethal mutations, miscarriage, sperm cell damage and immune deficiency have clear medical consequences.

The Indian government asked two respected scientists, Dr P.C. Kesavan and Dr P.V. Sukhatme, to investigate the possible reasons for differences in the results of the Bhabha Centre (BARC) and the National Institute of Nutrition (NIN) studies. At that stage both centres had published only preliminary findings. The BARC was permitted to submit additional data to support their conclusions. The NIN was not. Having read the report of this two-person committee we have to agree with Dr Srikantia, the director of the NIN, that,

> The committee appears to have sat more in judgement on NIN's work to discredit it, than try and determine the

> reasons for differences between NIN and BARC. We are therefore constrained to state that we feel the report is not an unbiased one.[72]

The reasons for the differences remain unresolved.

Various misleading and inaccurate claims have been made about the NIN studies.

1. It has been claimed that that the NIN study on polyploidy in children was *fraudulent and that it had been repudiated* by the director of the NIN.[73]

In fact, according to the Director, Dr S.G. Srikantia, the Institute fully supports the work of its researchers and has answered in full all criticisms made of this work.[74]

2. It has been claimed that the NIN experiments were *poorly designed* and consequently, their *results were imprecise*.[75]

In fact no one has been able to state what was wrong with the design. The same basic design was used by the BARC and by the International Food Irradiation Project. These other studies, as we will show, display gross deficiencies in conduct of the experiments that are not found in the NIN experiments or have major differences in the diet that may mean they are not really comparable studies. The NIN results were quite clear and far from imprecise. It is also hard to understand how Dr Sukhatme could have made this criticism. In the published reports he is acknowledged as having advised on the conduct of the experiments at the design stage.[76]

3. The *number of animals* studied is *small*. [77]

This is true if one considers each individual experiment but five health effects found in four animal species showed consistent results. The polyploidy studies were done on three separate occasions in rats and once in mice. Dominant lethal mutations were found in rats and mice. At the suggestion of Dr Sukhatme, the NIN undertook another study in mice which confirmed the original findings. Kesavan and Sukhatme then refused to consider the results, claiming that it was outside their terms of reference.

4. The observed increase in polyploidy is due to *small sample size* and/or to *faulty and biased selection* of the sample.

The initial number of cells counted was small. Subsequently the researchers undertook a total count of all polyploid cells on the slides. At all stages the slides were coded to remove the possibility of experimenter bias. The slides were also independently evaluated by another experienced cytogeneticist in India. These results, which Drs Kesavan and Sukhatme refused to consider in their report, have now been published.[78] They clearly confirm and strengthen the initial findings.

5. The maximum frequency of polyploidy found in the NIN studies is 1.8 per cent and this is *within the normal range* of incidence in polyploidy in normal healthy human beings[79] which can be up to 4 per cent[80] or 5 per cent.[81]

Drs Kesavan and Sukhatme claim that a normal level of 5 per cent is 'well documented'. In fact they are able to cite just one study on patients undergoing drug treatment where the average value was 2.5 per cent (range 0 per cent to 5 per cent). There were no healthy controls in this study for comparison.[82] On the other hand there are numerous studies which document normal incidence of polyploidy in the range of 0.06 per cent to 0.4 per cent[83] — most of them with values comparable to the 0.0 per cent to 0.17 per cent reported at the start of the NIN studies. We asked Peter Elias, the former head of the International Food Irradiation Project, what evidence there was for the figure of 4 per cent. His reply was that he telephoned Dr Evans, a leading cytogeneticist at Edinburgh University, and was told around 4 per cent.

We doubt that many scientists who have repeated this criticism of the NIN studies would have done so if they realized that it was based on no more than one study of unhealthy people and a single telephone call. If there is good evidence that the normal level of polyploidy is 1 per cent, 2 per cent, 4 per cent or 5 per cent, rather than the much lower figures cited by the NIN and other studies, we would like to see it. To put it in plain language, those who wish to use this argument should either 'put up or shut up'!

Whatever the normal level of polyploidy in a human population, each animal (and most of the children) showed an increase with continued feeding of freshly irradiated wheat followed by a fall when the irradiated wheat was withdrawn.

6. The NIN results are *theoretically impossible*.

Some criticisms strain scientific logic to the extreme. Dr Kesavan states,

> [The NIN data] raise many questions which *cannot* at present *be explained* in the light of well known biological principles and phenomena. It is yet another thing if biological principles are re-written so as to accommodate the NIN data.[84]

Similar statements have been repeated in Australia[85] and at the United Nations International Conference on Food Irradiation in 1988. In short, the NIN studies must be wrong because their data don't fit current theories. Many of the major advances in science have come when observations did not fit established theory. When fact does not fit theory should we follow the lead of these scientific experts and throw out the facts? We believe that there is important evidence here that requires explanation and if some cherished ideas have to bend to new facts then so be it.

7. The polyploidy observed by the NIN studies was only a *transient* effect of no medical significance.

The UK Advisory Committee illustrates its own peculiar brand of logic when it states,

> We noted the effects of one study in which irradiated wheat . . . was fed to malnourished children . . . It was found that the abnormal cells disappeared within a few weeks of withdrawal of irradiated wheat from the diet and we did not think that this transient phenomenon would have any harmful long-term consequences.[86]

In fact, the polyploidy in the study on children was observed after 6 weeks and the study was immediately terminated. In the animal studies the effects were observed again at 6 weeks and feeding continued to 12 weeks. In all cases it was observed that the level of polyploidy returned to normal a few months *after withdrawal of irradiated wheat from the diet*. How the Advisory Committee can conclude from this that there would be no medical consequences from continuing to feed irradiated food to a whole population on a long-term basis, we fail to understand. It did not convince the British Medical Association either. The BMA's Board of Science referred specifically to the NIN studies when concluding in 1987 that the Advisory Committee's advice had not fully taken account of possible medical consequences and that further research was needed.[87] The UK Advisory Committee appears to be unaware of the animal studies and the wider health concerns raised by the NIN research. It also reported that the irradiated wheat was fed 'at a dietary level of 50 per cent'. In fact, the level was 70 per cent. Just how thoroughly did the Committee review this evidence?

8. It has also been claimed that: the issues raised by the NIN research have been fully *discussed by the UN Expert Committee*;[88] the Indian government report has shown they can be *discounted*; that *no other studies have found the same effects*; and that *other studies have all shown that these effects do not occur.*

All the above are untrue. The UN Expert Committee merely noted the effects reported by NIN and did not comment on the Kesavan/Sukhatme report or the various claims and counter-claims made about these studies. Other studies, as we have shown, do confirm some of the NIN findings. As far as we are aware, there have been no other studies designed to investigate the observed immune deficiency reported in the NIN studies.

The International Food Irradiation Project study

In an attempt to resolve the controversy, the International Food Irradiation Project commissioned a major study in 1977.[89] This study was supposed to use the same methodology as the NIN polyploidy studies. A close look at the published report raises more doubts than are laid to rest. The experimenters (to their credit) report that on week 8 they could not find the feed for the experimental animals and concluded that they must have fed it to the control group. Rather than repeat the whole experiment the IFIP recommended that a new control group be fed for the full 12 weeks on unirradiated wheat, and polyploid cells from this group be compared with the experimental animals. Unfortunately this invalidated the careful matching of case and control animals which was done by the NIN researchers and built into the design of the IFIP study. Additionally, when two different researchers counted the number of polyploid cells they obtained very different results, even though they were from the same animals. This suggests possible further flaws in the implementation of the study. Even more important, the animal diet used was markedly different from that used by NIN.

Table 3 shows the dietary composition of the National Institute of Nutrition (NIN), Bhabha Atomic Research Centre (BARC), Osmania University (OU), and International Food Irradiation Project (IFIP) studies. The BARC, OU, and IFIP studies added additional ingredients and used different types of oils. These supplied extra nutrients — in particular, extra antioxidants and other free radical scavengers. For example, Bengal gram contains flavonoids which are strong antioxidants and increase the effectiveness of vitamin C as a free radical scavenger. Lentils help detoxification in the intestines. Sesame, shark and corn oil contain much higher levels of selenium and vitamin E than the groundnut oil used by NIN which was chosen because it is commonly used in India. Fish meal and dried yeast are also excellent sources of selenium and European wheat is also richer in selenium than Indian wheat. The IFIP study contained higher levels of other trace minerals such as magnesium, zinc, and manganese, which are essential to the free radical scavenging process. The IFIP also gave massive additional doses of vitamins A and D 'by gavage', that is, by tube directly into the animals' stomachs once a week. Even the wheat content was different and the diets were supplemented with additional gram flour, fish meal, groundnut cake, yeast powder, etc. The result was markedly different protein contents: 9 per cent NIN, 16 per cent BARC, 23 per cent OU, and 13 per cent IFIP. None of the other laboratories used the diet selected by the Indian National Institute of Nutrition as appropriate for testing the potential health effects of irradiation on the Indian population. These differences are critical. They could help explain the differences between the results of the various studies. They also indicate that the nutritional effects of irradiation may be much more significant than the expert committees suggest.[90]

Table 3: Ingredients used in the diets of conflicting studies on the safety of irradiated wheat[91]

Ingredient (gm/kg)	NIN[1]	BARC[2]	BARC[3]	OU[4]	IFIP[5]
Wheat flour	700	750	700	600	700
Bengal gram flour	—	—	200	80	—
Casein	—	60	—	—	150
Choline chloride	1	—	—	—	1
Corn oil	—	—	—	—	50
Fish meal	—	—	50	80	—
Groundnut cake	—	—	—	200	—
Groundnut oil	50	—	—	—	—
Sesame oil	—	60	2.5	—	—
Shark liver oil	—	—	7.5	—	—
Starch	200	—	—	—	50
Sugar	—	90	—	—	—
Yeast powder	—	—	40	30	—
Mineral mix	40	20	—	10	40
Vitamin mix	10	20	—	traces	10
Weekly supplements					
vitamin A	—	—	—	—	1000 U
vitamin D	—	—	—	—	100 U
Protein content (%)	9	16	16	23	13.2
Weeks of feeding	12	6	1-12	24-52	12

1: *Vijayalaxmi and Sadasivan, G., Int. J. Radiat. Biol., 27, 135-142, 1975.*
2: *George, K. P. et al., Food Cosmet. Toxicol., 14, 289-291, 1976.*
3: *Chauhan, P. S. et al., Toxicology, 7, 85-97, 1977.*
4: *Reddi, O. S. et al., Int. J. Radiat. Biol., 31, 589-601, 1977.*
5: *Tesh, J. M. et al., IFIP R45, 1-64, 1977.*

Studies on human volunteers

There have been several studies in the USA and China on the effects of feeding irradiated food to humans. Conscientious objectors were fed irradiated foods for two weeks in a US university study in 1953. No immediate damage was observed but no long-term follow-up was built into this research.[92] The US army also performed human feeding trials. There is little information in the scientific literature on these studies. Chinese researchers reported in 1982 that they were 'unable to get the information in detail'.[93]

Most of the work on feeding irradiated foods to human volunteers has been done in China. Reports on these have been reassuring. Studies on feeding potatoes,[94] rice,[95] mushrooms,[96] meat products[97] and peanuts[98] all found no adverse effects. These were followed by eight human volunteer studies in 1982. The studies investigated rice, mushrooms, sausage, peanuts, meat products and potatoes. There were also two studies where 60 per cent to 66 per cent of the whole diet was irradiated. These extended over 7-15 weeks and looked at a variety of health factors. In no case was any difference found between the volunteers in the experimental group and those in the control group who were on a normal unirradiated diet.

A subsequent human volunteer study[99] in 1984 tested the effects of irradiating the whole diet consisting of 35 different foods. This involved 70 volunteers and lasted for 90 days. The only possible adverse effect noted was a small increase in polyploidy. This was not thought to be significant as a similar increase in polyploidy was also found in the control group. However, the details of this study were initially given only in the 1986 proceedings of an International Atomic Energy Agency seminar. A book promoting food irradiation, published by the World Health Organization, suggests these Chinese studies are the last word on the subject. It states that,

> human volunteers in China . . . showed no signs of any health effects, including chromosomal changes.[100]

This is one of the more irresponsible statements in the WHO book which has been criticized by the International Organization of Consumers' Unions as biased, misleading and inaccurate. The details of these Chinese studies, when they were finally published in full in the open literature, show that the food for the volunteers was *stored*, in some cases for up to four months, before being consumed.[101] Far from disproving the findings of the NIN, they confirm one major finding — provided food is stored for a sufficient time after irradiation, these adverse health effects are not observed. They say nothing on the crucial question of freshly irradiated food. This remains an open question.

Is irradiated wheat a hazard?

Clearly, if there is a problem with some freshly irradiated foods these may need additional controls. Irradiated wheat may have to be labelled with a 'do not sell before' or 'first date for sale' date to prevent its consumption within whatever time period is considered appropriate. This was a recommendation of a Consensus Conference initiated by the Danish parliament in 1989.[102] The question of the safety of irradiated wheat was discussed in some depth at this conference which was attended by both Dr Vijayalaxmi, one of the NIN researchers, and Dr Elias, former head of the IFIP. When we asked Peter Elias for his comments on the conflicting claims and counter-claims that had been made for the safety of freshly irradiated wheat, he conceded that there was a need to review the evidence in the light of the new information provided by Dr Vijayalaxmi. However, he maintained that the issue did not represent a significant health problem — because no one would be eating freshly irradiated wheat anyway. His view was that it would always be stored before consumption. We suggested that this was not the point. The official view was that there were no safety problems, however long the food was stored. He insisted that no one was claiming this and asked us to provide evidence. We handed him the report of the US Council on Agricultural Science and Technology which quoted from the foreword to the IFIP report on polyploidy:

> The results of these studies showed that, in contrast to the Indian findings, neither the incidence of polyploidy nor the incidence of micronucleated cells were affected significantly by a diet containing flour prepared from irradiated wheat *irrespective of the time of storage*.[103] [emphasis added]

The foreword to the IFIP study was written by Peter Elias in 1977!

It was noticeable that Peter Elias did not challenge the final recommendation of the conference panel that there was a need for mandatory storage of some irradiated foods before consumption.

Incompetence or fraud?

It is unclear whether the various scientific experts are unable to put together the facts that suggest there may indeed be a biochemical — nutritional mechanism for harm from eating irradiated food, or whether there has been a deliberate attempt to cover it up. The following factors are significant:

- The UK Advisory Committee acknowledges that irradiation produces potentially hazardous free radicals which initiate chain reactions leading to peroxidation and epoxidation in polyunsaturated fats. It also acknowledges that this process may continue for several weeks after irradiation.[104]
- The UN Expert Committee report in 1977 suggests that there was a need for 'further chemical, nutritional and toxicological studies on the radiolytic products of lipids (fats) with reference to peroxide and epoxide formation.'[105]

These studies, which promoters of irradiation claim have been done,[106] were not reviewed by the UN Expert Committee in 1980.[107]

- Methods for discovering whether some foods or food components with low water content have been irradiated rely on detection of residual free radicals.[108]
- In its second report, the UK Advisory Committee acknowledges that irradiation may require the additional use of 'antioxidants'.[109] The natural antioxidants in food are the vitamins A, C and E and some of the B vitamins especially in combination with other food components and trace minerals.
- These vitamins, as we will show in the next chapter, are severely damaged by irradiation.
- These same vitamins and some trace minerals also appear to have been added to the diet of animals in some experiments that are used to suggest irradiated food is safe.

This issue of vitamin supplementation does not apply solely to the studies investigating chromosome damage. The World Health Organization states,

> more than 100 generations of sensitive laboratory animals in the United Kingdom alone have been living and prospering on diets sterilized by irradiation.[110]

We have no information on the diets these UK animals were fed, but a similar claim was made about animals at the Melbourne-based Walter and Eliza Hall Institute during the Australian Parliamentary hearings into food irradiation.[111] Our colleagues in Australia made an analysis of the diets of these animals and found, in some cases, levels of vitamin supplementation 6-36 times higher than normal.[112]

It is acknowledged that problems with blood clotting which were reported in early studies were attributable to irradiation damage to vitamin K.[113] [114] It was necessary to add vitamin K supplements to the diet to overcome the deficiency caused by irradiation. The UK Advisory Committee suggests that this loss of vitamin K may be a significant issue for pregnant women and nursing mothers.[115]

To what extent are there vitamin and other dietary differences between the various studies showing results for and against the safety of irradiated food? Clearly one major difference between the studies on polyploidy was the level of vitamin, mineral and protein supplementation. How many of the other experiments also used vitamin supplements, at what levels and why?

The answer we have been given to these questions is that vitamin supplementation is routine in toxicological experimentation as one is seeking to identify the effects of exposure to the radiolytic chemicals and does not want these confused with nutritional deficiencies.[116] However, while scientific investigation may wish to deal with the issue in neat compartments, the public needs the issues of safety dealt with as a whole. Either the various scientific experts are unaware of the way these issues of toxicology and nutrition are interconnected or there has been a systematic attempt to confuse the issue and cover up potential adverse health effects. Which are we dealing with, incompetence or fraud?

Moving the goal posts

Earlier in this chapter we identified various relaxations in the requirements for testing of irradiated foods. For all the assurances, irradiation has not been tested to the same degree of stringency as would be required for additives to food. Tracing the history of safety testing of irradiated foods is like watching a football game in which the referee is continually moving the goal posts — in favour of one team.

- Initially, irradiation was defined as an additive; then, when the problems of testing it as such became apparent, it was re-defined as a process to be tested under an entirely different regime.
- The UN Expert Committee initially required testing of all food products. In 1976 this requirement was removed. Results from one food could be applied to another of a similar type. For example, data on safety testing of wheat can be applied to other grains. This is despite the view strongly expressed in the Hungarian Academy of Sciences report that the data from the studies on the safety of different food products is on the whole ambiguous and unreproducible — that data from study of one food cannot be applied to another.[117]
- Studies using doses above 10 kGy do not yet provide convincing evidence of safety. The expert committees insist that the clearance of irradiation up to 10 kGy should not be interpreted as indicating irradiation is unsafe above this level. Yet when we cite studies indicating adverse health effects, these are dismissed if the dose applied is above 10 kGy.
- Clearly, if there is a safety threshold then control of the irradiation dose is critical. Initially the UN Expert Committee specified both maximum and minimum doses that should be used on particular foods. The minimum dose guaranteed that the food would have the changes expected of it and the maximum guaranteed that undesirable effects would be limited. In 1981 this requirement was removed and only the average dose specified.[118] In practice the maximum dose can often be two or three times the minimum. The 10 kGy average is therefore effectively permitting doses up to 15 kGy for some parts of a food

batch.[119] This change appears to have more to do with commercial considerations than with public health.

- Studies which found safety problems are dismissed if other studies do not find the problem. Greater priority has been given to discrediting studies indicating problems than to finding the reason for conflicting results. In particular, health problems reported in some early animal studies appear to have been remedied by altering the diet — increasing the level of vitamin supplementation. Later studies which do not show the problem fail to acknowledge that this may be the reason but, instead, claim that they prove there are no safety problems.
- Studies which find adverse health effects are also dismissed on the basis of supposed flaws in their methodology. Yet when the US Food and Drug Administration applied the same criteria for acceptability to 400 different studies, for and against the safety of irradiation, it was left with only five on which to base approval of irradiation. When some of these five also suggested there might be problems, the FDA argued that irradiation could be approved (though only up to 1kGy) on the basis that the amount of chemical change caused by irradiation is so small as to pose only an acceptable risk. This has been described by our US colleagues as rather like taking the goal posts away completely.
- Early expert committee reports on the safety of irradiated food, such as that in 1964 which led to the process being banned in the UK in 1967, were fully referenced. Later expert committees seem unable or unwilling to provide comprehensive referenced reports. The reports on which the official assurances are based appear to contain statements which are biased, illogical, misleading or inaccurate. Some claims are made which just cannot be supported by the facts.

The whole area of testing and reviewing the safety of irradiated food needs to be re-opened to public scrutiny. The public has a right to unbiased objective information on the possible harmful effects of irradiation and not bland reassurances that hide uncomfortable evidence under value judgements. It is the job of the democratic processes to decide whether the effects are significant and the risks acceptable — and to lay down the conditions under which any risks can be minimized. When science and scientific experts enter this area to pre-empt discussion of these issues we have reason to be concerned. The World Health Organization must be asked to re-open the investigation and produce a comprehensive, properly referenced report. Without this, we suggest, the public would be most unwise to accept the national and international assurances on safety that are used by promoters of irradiation.

4. Is it good to eat?

The wholesomeness of irradiated food

In the English language, wholesome is a word that combines the ideas of safe, nourishing and health-promoting. There is no similar word in other languages. This is unfortunate. It means that, in the international debate about food irradiation, the word has been debased so that it really only covers the absence of any harmful effects. As long as there are no toxic chemicals in the food, or the microbiological and nutritional changes do not cause obvious problems, the expert scientific committees feel able to declare irradiated food safe and 'wholesome.'

It may be possible to show that, properly controlled, irradiation will not produce particular harmful agents, or that the risk of such events is acceptably small. Even so, the term wholesome cannot truthfully be applied to irradiated foods. They may still appear fresh but this is counterfeit freshness. They may have undergone significant losses in essential nutrients either by the process of irradiation itself or from the extended storage that irradiation makes possible. There is a big difference between 'safe for human consumption' and 'wholesome'. This is a distinction the public can and does make, and has a right to continue to make.

This right can only be guaranteed by honest labelling so that the consumer can choose between fresh, wholesome food and processed, irradiated food. Wholesomeness and consumer information are issues that cannot be separated. But it goes deeper than protecting the individual's right to know what is being done to food. There is also a need to consider the effects on public health if a significant portion of the diet comes from irradiated foods.

Nutritional effects of irradiation

The effects of irradiation on the major food components are small. The dietary value of protein is only slightly affected. Many of the changes merely break down proteins in the same way as occurs during digestion in the body, although the nutritional usefulness of some components of proteins may be reduced.[1] The dietary value of starches and sugars is also little affected. The main concern with carbohydrate foods is the potentially toxic effects of chemical changes leading to such radiolytic products as formaldehyde.

Fats, however, do not take kindly to irradiation. They develop rancid flavours and odours when irradiated. The changes in fats and in sulphur-containing proteins have been identified as the source of most of the undesirable changes in irradiated foods that lead, for example, to:

- meat developing what has been described as a 'wet dog' smell[2] [3]
- fats tasting 'nutty' or 'oily'[4]
- milk and most milk products tasting 'like burnt wool', 'scorched', 'candle-like' or 'chalky'.[5]

The problem with fats is so severe that most milk products and oily or fatty foods are not even being considered for irradiation.

There is a growing consciousness among many people of the need to make changes in diet on health grounds — not least to reduce the appalling toll of coronary heart disease. Government health agencies and the World Health Organization advise reducing overall fat consumption and substituting unsaturated fats for the saturated fats commonly found in animal products.[6] Alongside this change towards greater use of polyunsaturated fats and oils, there is also a recognition of the vital role of some essential polyunsaturated fatty acids in promoting health. Polyunsaturates are damaged even more rapidly by irradiation than saturated fats. There has been little research into this aspect of dietary health and the consequences for public health need to be investigated.

Indeed, this concern over lack of research can be applied more generally. The very foods that we are being encouraged to eat — white meats such as chicken and fish in preference to red meats, whole grains in preference to highly processed starches, and more fresh fruit and vegetables — are the very foods that will be targeted for irradiation. All of these will undergo some degree of damage to essential nutrients.

Vitamins

This is of particular concern in the case of vitamins. It is not in dispute that irradiation causes damage to many vitamins. Vitamin A and its precursor beta-carotene, vitamins B1 (thiamine), B2 (riboflavin), B3 (niacin), B6, B12, folic acid, vitamin C (ascorbic acid) and vitamins E and K are all damaged to a greater or lesser extent by irradiation. Vitamin E, which is known to protect the polyunsaturated fatty acids, is so badly damaged that in many cases it is even destroyed if it is put back into the food as an additive after irradiation.[7]

Table 4 indicates some of the research findings on vitamin losses from irradiation. Some of the figures (indicated *) are losses resulting from irradiation at higher doses than currently being considered. The extent of the loss depends on the dose, the vitamin and the type of food. Generally speaking, the more complex the food, the less it suffers vitamin losses during irradiation. Fruit juices will suffer more than fresh fruits and these more than vegetables, grains and meat products. Nevertheless, as the table shows, losses of 20-80 per cent are not uncommon and there are still many gaps in the available scientific data on vitamin losses.[8] [9] [10] [11] There has been no serious assessment of the impact on public health if a significant portion of the population's diet were to be irradiated, even at the 'lower' doses currently recommended.

Even quite small losses can be significant for some groups within the population such as the elderly, the young, the sick and the poor, many of whose diets are already deficient. In the UK, a Department of Health study found that many school children suffer some form of dietary deficiency.[12] Official reports in the USA[13] and in Australia[14] also show that there are groups of people whose diet is already deficient, particularly those whose diet is limited by low income. The Australian survey found between 16 per cent and 33 per cent of the population had vitamin intakes below the recommended daily

Table 4: Some reported vitamin losses from irradiation[2, 9, 10, 21, 29]

	Vitamin							
Food	A	B1	B2	B3	B6	B12	C	E
Milk	60-70	35-85	24-74	33	15-21	31-33		40-60
Butter	51-78							
Cheese	32-47							
Grains and flour								
Wheat	—	20-63	—	15	3	—	—	—
Oats		35-86						7-45
Rice		22						
Beans	—	—	48	—	48	—	—	—
Meats								
Beef	43-76*	42*-84*	8-17*	—	21*-25*	—	—	—
Pork and ham	18*	96*	2*	15*	10-45*			
Chicken	53-95*	46-93*	35*-38*		32*-37*			
Eggs	—	24-61*	—	18	—	—	—	17
Fish								
Cod	—	47	2*	—	—	—	—	—
Haddock		70*-90*	4*					
Mackerel		15-85*			26			
Shrimp	2-27	70*-90*						
Potatoes	—	—	—	—	—	—	15-56	—
Fruits and vegetables							30	
Fruit juices	—	—	—	—	—	—	20-70	—
Tomatoes							14	
Nuts	—	—	—	—	—	—	—	19-32

Note: some vitamins are relatively undamaged by radiation but absence of a figure in the table above does not imply that a food has been cleared. In general more work needs to be done on a comprehensive study of vitamin losses.

All percentage losses are at doses below the 10 kGy proposed clearance level unless *. In these cases doses are below 60 kGy being suggested for sterilization of meat products.

allowance. Bakers are encouraged to add thiamine (vitamin B1) to flour 'to improve the thiamine status of Australians'.[15] A similar pattern of deficiency in susceptible groups exists in most developed countries and would apply, to a much greater extent, in less developed countries.

In developing countries, vitamin A deficiency is a major preventable cause of blindness (xerophthalmia). It occurs against a background of poverty and food shortages. If the already marginal food supply in these countries were to be irradiated for longer storage, the losses in vitamin A could result in this already common disease affecting even more people.

The average European diet contains only about half the vitamin C that the USA recommends as the daily allowance. When there is not enough fresh fruit in the diet, potatoes are the major source of vitamin C. If potatoes are irradiated, as has been suggested, the amount of vitamin C in the diet could be significantly reduced, possibly compromising the health of those with already marginal intakes. Any further reduction in vitamin C for those with marginal intakes is not desirable.

Folic acid (folacin, folate) is an essential vitamin from the group of B vitamins. Deficiencies in folate and of other vitamins[16] and minerals[17] have been linked to the development of neural tube defects such as spina bifida. Even with a normal diet, pregnant women, young children, the elderly and alcohol abusers are particularly at risk of folacin deficiency. In addition, studies of elderly people and people entering mental hospitals have found a significant proportion of such people with deficiencies in folic acid. In its brief and inadequate discussion of the issue, the UK Advisory Committee states that:

> little is known about the effects of irradiation on folate. Since there are possible public health problems in relation to the intake of folate this needs further investigation.[18]

Are these losses significant?

Unfortunately, food irradiation advocates see nothing but benefits from irradiation in the war on hunger and malnutrition. Their dream rapidly turns into a nightmare if we think of irradiation being used to extend the storage life of the developed world's surplus food mountains so that these nutritionally depleted stockpiles can be off-loaded onto third world countries — adding insult to injury by calling it 'aid'. In developed countries, irradiation will not be used on the premium quality produce. This already commands the highest price in the market-place. Its use on lower quality food to provide the illusion of freshness will have greater significance, in nutritional terms, for those who may already be living on marginal diets.

In the previous chapter we showed how the presence of free radicals and radiolytic chemicals combined with the loss of essential nutrients could have significant health effects. These effects were found in studies which used the 'adequate but marginal' diet selected by the Indian National Institute of Nutrition. In studies by other institutes, where the diet was fortified with extra vitamins, the effects were not observed. Why? It is quite possible that the extra vitamins masked the effects. They were the very antioxidant vitamins needed to neutralize the free radical chain reactions initiated by irradiation.

Table 5: Average vitamin losses during household preparation of foods[19]

Vitamin	Per cent
Thiamine (B1)	30
Riboflavin (B2)	15
Niacin (B3)	20
Ascorbic acid (C)	35
Folic acid	40-50

Far from recognizing that there might be a problem, the response of the pro-irradiation lobby again illustrates our concern over the way such public health matters are dismissed. We are told that the vitamin losses caused by irradiation are not significant because they are comparable to those occurring naturally in cooking, storing and other forms of food processing. This is both untrue and misleading.

- Many of the foods intended for irradiation — such as fruit and vegetables — are those that would be eaten fresh and often raw. Comparisons with cooking and storage just do not apply. Irradiation losses are additional to any cooking and storage losses, which can in any case be considerable, as shown in Table 5.

- Some vitamins — B1 and E for example — undergo accelerated losses during storage after irradiation.[20] An example of the combined effect of irradiation and storage is given in Table 6. This shows that the losses of vitamins B1 and E in wheat and oats can be three times greater with irradiation. Losses of 67 per cent and 85 per cent, compared to 25 per cent for the unirradiated grains, cannot be considered insignificant for populations heavily dependent on such staple foods.
- The object of irradiation is to permit increased storage times while retaining the appearance of relative freshness, so storage losses will be increased beyond those occurring normally.

Table 6: Vitamin losses during storage (8 months) increased by prior irradiation[21]

	Vitamin losses caused by irradiation	Vitamin losses caused by storage	Total	Non-irradiated control losses caused by storage
Vitamin B1 in wheat flour Dose 0.25 kGy	20	47	67	25
Vitamin E in rolled oats Dose 1 kGy	20	65	85	26

The food can thus undergo initial losses on irradiation, accelerated losses during storage, additional losses because of longer storage times, and then lose further vitamins in cooking. What this adds up to is poor quality food. For consumers who are increasingly conscious about the quality and vitamin content of fresh foods, irradiation counterfeits freshness and conceals what may be significant losses of essential health-giving nutrients.

In Britain, where there has been a furious debate on the issue of vitamin losses, the irradiation lobby has used a fall-back argument when confronted with these facts. Sir Arnold Burgen, the chair of the UK Advisory Committee, said that it would not be a problem because no one was likely to eat a significant portion of their diet from irradiated foods.[22] This comes dangerously close to saying irradiated food is all right as long as you don't eat it!

At an earlier meeting with the British Food and Drink Federation, an industry spokesperson challenged us to name any food that was likely to be both irradiated and consumed in significant quantity. We suggested that the British were known world-wide for their consumption of fish and chips — and there was a rapid change of subject. Given that irradiation is considered appropriate for fish, chicken, potatoes and other vegetables as well as for grains and some fresh fruits, it is clear that some very important staple foods could be affected.

Faced with widespread public concern, the British Advisory Committee has suggested that there should be long-term monitoring of irradiated food for nutritional damage.[23] It would, we suggest, be better if this research were done before the widespread application of irradiation to food, not after. As with the area of food safety, the opinions of the expert committees are no substitute for scientific evidence. In the absence of hard evidence on nutritional changes and their impact on people on marginal diets, the widespread consumption of irradiated food would be an uncontrolled human experiment.

Labelling — a public health issue?

Irradiated foods will look fresh for longer. The consumer will be encouraged to view these as healthy and wholesome whereas they are likely to be older and of lower nutritional value. In these circumstances there is considerable potential for deceit — for what can best be described as 'counterfeiting' fresh food.

In many surveys that have been done around the world,[24] consumers have consistently demanded that all irradiated food be labelled. Some simply desire to know how the food has been treated so that they are not misled. Some clearly wish to avoid irradiated foods. Some may see benefits in it for which they are prepared to pay more. And some will need to know so that they do not, in fact, end up having too great a portion of the diet irradiated — at least until the full dietary impact of irradiation has been adequately assessed.

Even if, as its promoters claim, irradiation were to cause nutritional losses merely comparable to those in other forms of food processing, then consumers should regard irradiated food in the same way as food taken from the freezer, the packet or the can — food that might be eaten for convenience but not for health. Since government nutritional advice now encourages the public to balance processed food with fresh food, the consumer will need to balance irradiated food with fresh food. The problem is that irradiated food looks like fresh food. It looks like fresh food for longer, even though it will be nutritionally depleted even before storage and even more so if irradiation is used to extend its 'shelf-life'. Consequently, labelling of irradiated food is not merely an issue of choice — the consumers' right to know; it becomes a public health issue — the consumers' need to know in order to be able to provide a healthy balanced diet.

In the UK the government has agreed that all irradiated foods and irradiated ingredients should be labelled. The United Nations CODEX Alimentarius Committee and the European Commission have also conceded this point. Faced with united demands from consumer organizations world-wide, they could do little else. However, exactly how this will apply to foods sold loose and in bulk remains to be seen. We also wait to see if irradiated food sold through catering outlets, such as restaurants and canteens, will be labelled. We suspect not.

Not all countries have agreed to honest labelling. Chinese government officials have declared that irradiated food will be labelled when there is consumer acceptance and not before.[25] In the USA, there has also been considerable reluctance to compel labelling of all irradiated foods. At one stage there was vehement opposition to any form of labelling. Legislation was introduced in both Houses of the US Congress to prevent individual States from enacting their own legislation to require labelling. In 1984, Margaret Heckler, Secretary of Health and Human Services under the Reagan administration, first tried to delete the Food and Drug Administration's proposals for limited labelling and then suggested that the label should contain the term 'picowaved' — hoping no doubt to draw on the widespread acceptance of 'microwave' technology, even though microwaves cannot alter the food chemistry as ionizing radiation can. Fortunately, within the corridors of power this idea 'didn't pass the laugh test'. The USA currently requires labelling with terms such as 'treated with ionizing radiation' and 'irradiated to control . . .' However, there is no requirement to label ingredients. A food could be made up largely of irradiated ingredients such as spices, wheat flour, dried vegetables etc., and not have to be labelled. The requirement to label any irradiated food with words may lapse in 1990, allowing US food manufacturers to use only a symbol.

This symbol (shown in Figure 3), we are told, is the international symbol for irradiated food, originally developed in the Netherlands. It is used in South Africa for this purpose but up to 1990 we were unable to find it on any food product on sale in the Netherlands.[26] Since the Dutch irradiation company has claimed that all the food irradiated was 'for export' this is possible. However, we have been unable to find it in use anywhere else in Europe either. It seems that the symbol is being promoted in the hope that it will *become* the accepted international symbol — another example of myth rather than reality in the promotion of irradiation.

Symbols alone are inadequate. They can mislead rather than inform. Consumers will need to ensure that all irradiated food is clearly labelled with the term 'irradiated'. Consumers may also wish to see more information provided. There are strong pressures for more complete disclosure of nutritional

Figure 3: The Radura symbol — being promoted as the international symbol for irradiated food

information on the labels of processed foods. It is unclear how the food industry will assess the true vitamin content of irradiated food if such labelling is required. Consumers may also seek to know how old the food is. Again it is unclear what this will mean for the current 'best by' and 'sell by' (or the more international 'use by') date marking system. If the findings of possible effects from freshly irradiated foods are correct it may also be necessary to have a 'do not sell before' date on some foods. This proposal was recommended by a consensus conference established by the Danish parliament in 1989.[27]

Detection methods

Given the food industry's reluctance to label irradiated products, how will labelling requirements be enforced? We raised this question in 1985. Up to that point there had been almost no attention given to the question of controls. In the 70 years of experimentation that promoters claim has gone into irradiation, it seems that almost no one thought to do the research that would enable irradiation to be detected. There are still no appropriate tests that will allow the agencies responsible for consumer protection to determine whether any food has been irradiated and if so how many times and with what doses.

Without detection tests, the system is wide open to fraud and abuse. Look at it this way: if, for the sake of argument, we assume that irradiated food is safe, wholesome and nutritious, that consumers are prepared to accept it and even to pay more for it, then what is there to stop unscrupulous traders labelling food as 'irradiated' and charging a higher price without it ever being near an irradiation plant? On the other side of the argument, what incentive is there to label as 'irradiated', food which consumers might reject? Is the whole system of controls to depend on honesty in the food trade? Most countries enforce an elaborate system of food regulations simply because it is unrealistic to expect everyone involved to be honest.

In 1987 the UK government decided to begin financing research into methods of detection. Efforts are currently concentrated on four techniques.

- *Electron spin resonance* — a spin-off from the medical technique of magnetic resonance imaging — detects the presence of residual free radicals. This may be appropriate for dry foods such as grains and spices or those with bone or shells which retain the free radicals for longer.
- *Chemi-luminescence and thermo-luminescence* where irradiation is detected by light given off when irradiated food is heated or chemically treated. This technique might be appropriate for detecting irradiation of spices and dried products.
- Detecting *changes in the genetic structure* of food material.
- Techniques which detect *changes in food chemistry* especially proteins. These appear to be the most promising as the technology for analysing chemical composition of materials is relatively cheap and widely available in public analytical laboratories.

None of these has yet been developed to the stage of a cheap test which can be used routinely by public health and food law enforcement officers. It is not yet possible to verify that the controls suggested by the UN Committees have been adhered to.

Food quality — food prices

Will there be any economic advantage to the consumer? Will irradiation lead to cheaper food? The short answer is 'maybe, but at a cost'. There is no economic advantage to a food processor in irradiating the best quality food and thus having to pass on the additional cost of irradiation. Premium food already commands premium price. There is, however, considerable economic advantage to irradiating lower quality food so that it stays looking fresh for longer — like the better quality food it is counterfeiting. This can then be sold so as to marginally undercut the quality product. This is no hypothetical concern. Quality herb and spice importers in Europe are already concerned that their competitors will use irradiation to 'clean up' low-quality spices from southern Europe and North Africa, and so undercut the high-quality product. The whole spice market may soon be forced to accept irradiation — and lower quality standards — in order to remain competitive. Does anyone benefit from this?

These issues of labelling and enforcement are much wider than just ensuring the right of individual consumers to know that food has been irradiated. Knowledge gives some protection from manipulation by powerful companies in the food industry. These companies can manipulate prices, influence public perceptions through advertising, and conceal what has really been done to the food. In the final analysis, labelling and its enforcement give the consumer choices and some influence on food quality. The public needs to be assured that, if irradiation is to be used, there will be full disclosure of relevant information and the maintenance of the highest quality standards.

Above all, irradiation conflicts with the message that public health authorities — including the World Health Organization[28] — are promoting. Good quality, fresh food is needed to improve health. Irradiation reduces the health value while counterfeiting freshness. Irradiation is being marketed as suitable for use on precisely those foods which consumers are being encouraged to eat more of to combat coronary heart disease and food-related cancers: fresh fruit and vegetables, fish, seafoods, poultry and grains. By irradiating these foods, not only good foods but also health promotion are undermined.

5. Irradiation in the real world: can it be controlled?

There is no wholly risk-free technology. Most require a system of controls to ensure that society obtains the maximum benefit and is exposed to the minimum risk. In this chapter we will examine what has actually happened with irradiation in the real world, the problems of control over its use in the international food trade, and the risks to workers and the environment. We will also take a more critical look at some of the supposed benefits of irradiation, for example, its use in:

- dealing with food poisoning bacteria and insect infestation
- inhibiting sprouting of vegetables and delaying the ripening of fruits
- substituting for some other, possibly harmful, chemical preservation techniques.

On the surface these appear to be benefits. In the real world they may turn out to be of little or no benefit to consumers. Are there any benefits to the food industry and, if so, what are they? Is it the honest or the unscrupulous who stand to benefit most ?

Abuse of irradiation — the 'Dutching' of food

Late in 1985, environmental health officers alerted us to the possibility of illegal imports of irradiated food coming into the UK. We also received reports from within the food trade of widespread abuse of irradiation by companies in Europe —

using it to hide unacceptable levels of bacterial contamination on unsaleable foods in order to put these foods back on the market. Since then, a number of scandals have been exposed. We have discovered that, contrary to the assurances offered by its promoters, irradiation is not only wide open to abuse but such abuse is already occurring. Indeed, this is a major use for irradiation and it is only half-heartedly condemned by governments and the international promoters of irradiation. Furthermore, these abuses are almost impossible to eradicate since there are no reliable methods for detecting whether a food has been irradiated, and if so, how many times and with what doses. Far from offering a treatment process which can make food safer, irradiation has made obsolete many of the methods by which the public health and food law enforcement agencies can guarantee that food is safe, wholesome and fit to eat.

Youngs Seafoods (Admiral label) — contaminated prawns (UK)

The first case to come to light involved a consignment of Malaysian prawns originally imported into the UK by Young's Seafoods (now part of United Biscuits). Young's found these prawns were contaminated with a level of bacteria that made them unsaleable according to the company's standards. The prawns were shipped to the Gammaster irradiation plant at Ede in the Netherlands, irradiated, illegally re-imported into Britain and sold under the 'Admiral' label.[1] The Channel 4 consumer programme *4 What It's Worth* featured the case in April 1986,[2] two days before the publication of the UK government's Advisory Committee report.[3] It was also the subject of an Early Day Motion on the House of Commons Order Paper[4] (see Appendix B2).

Flying Goose (Dan-Maid label) — contaminated prawns (Sweden)

The TV programme also reported a similar case involving the Flying Goose company, part of the international Allied Lyons group. The company had irradiated prawns to conceal bacterial contamination using the IRE plant at Fleurus, Belgium, before shipping them to Sweden. In Sweden, as in Britain, the importation of irradiated food is prohibited. The company was reported as saying that the shipment had been a mistake and the practice would not continue. The consignment was rejected by the customer in Sweden following a tip-off and permitted to be re-exported.[5] The Flying Goose company is believed to have sold it legally as a 'reject' consignment. Its final destination is unknown,[6] but reports from within the food trade indicate that it changed hands several times and may have been repackaged to remove the original brand labels. Details of these cases were given to the UK government in 1986 but no action was taken.[7]

Hank DeBruijne and Highploeg — traders in 'reject' seafoods — and the Gammaster irradiation plant (the Netherlands)

There appears to be a thriving trade in the re-sale of foods which have been rejected as unsaleable for human consumption because of bacterial contamination. A TV programme named two Dutch companies, Hank DeBruijne and Highploeg, who regularly deal in 'reject consignments' and who do business with the Gammaster plant.[8]

Roemoe Museli — contaminated mussels (Denmark)

The Dutch parent company of a Danish company, Roemoe Museli, was convicted and fined in April 1988 for importing a 32-ton consignment of irradiated mussels. These mussels had previously been rejected by Danish authorities because of contamination with *E. coli* bacteria. The company had them irradiated at the Gammaster plant in the Netherlands and illegally re-imported into Denmark.[9] It is worth noting that no legal offence was committed by Gammaster. Even though the Netherlands does not permit irradiation of mussels for Dutch consumers, the product was being irradiated for export! It is, however, embarrassing for Gammaster which is on record as claiming that food is tested for bacterial contamination before being irradiated.[10]

International insurance fraud?

The International Maritime Bureau is investigating cases of possible insurance fraud where seafoods and frogs' legs may have been over-insured before being rejected by the authorities in the USA. It has been alleged that after insurance claims are made, the consignments are bought back as 'reject lots' and shipped to Europe for irradiation before being put back on the market. One consignment of frogs' legs is suspected of having crossed the Atlantic 11 times.

Na-Kagami Shokuhin — illegally irradiated baby foods (Japan)

In Japan, the Na-Kagami Shokuhin company, a subcontractor to such large Japanese food processing companies as Wakodo, Meiji and Snow Brand, was convicted in 1984 for irradiating vegetables for baby foods in containers labelled as 'animal feedstuffs'. The company president was sentenced to eight months imprisonment. The practice had been going on for four years between 1974 and 1978 and only came to light when one of the drivers reported how the boxes had been re-labelled as 'vegetables for babyfood' at drive-in restaurants on the way back from the Radiye Kogyo irradiation plant.[11] [12]

Illegal imports of irradiated spices (West Germany, UK and others)

It appears that these are far from being isolated cases. Many spices are irradiated and sold, unlabelled, in countries where it is still illegal, such as West Germany. UK Ministry officials have told us, privately, that some imports of pepper are known to be or suspected of being irradiated.

Quaker Oats (Golden Grain label) — over-irradiated mushrooms (USA)

In countries where spice irradiation is permitted, some companies have attempted to irradiate dried vegetables under these permits. In the USA, for example, spices are permitted to receive 30 kGy as opposed to 1 kGy for vegetables. In 1988, the Quaker Oats company was found to have violated the regulations by illegally importing irradiated dried mushrooms for use in its Golden Grain Chicken and Mushroom Rice-a-Roni and Chicken and Mushroom Noodle-Roni products. The mushrooms originally came from Taiwan. They were imported by the Cade-Grayson company in California and had been given ten times the dose permitted for vegetables in the USA at the plant of Precision Materials Corporation of Mine Hill, New Jersey. Quaker Oats/Golden Grain responded to widespread expression of consumer concern and discontinued the use of irradiated mushrooms.[13]

The 'Dutching' of contaminated seafoods — the practice continues (UK/USA)

The involvement of companies in the Netherlands in many of the cases of abuse that have come to light has led to the term 'Dutching' in the international food trade to describe concealment of contamination by irradiation.

In June 1989 a UK journalist, posing as a representative of a non-existent seafood company, showed how easy it would be to import irradiated seafoods from Hank DeBruijne and Gammaster. The seafoods originated in Malaysia but produce from Oman, Bangladesh, Pakistan and Taiwan was also offered. The results of the newspaper investigation were given in a House of Commons motion[14] (see Appendix B3). This motion names DeBruijne, Gammaster, a haulage company recommended by Gammaster as willing to import such foods into the UK, and three UK companies who trade with the Dutch companies.

The investigation highlighted a further problem. Port health authorities may not be able to detect whether a food has been irradiated but their suspicions are aroused if a food comes in with an abnormally low bacterial load. This can result in embarrassing and time-consuming questions for the importers. A Dutch journalist reported a conversation with the DeBruijne company in which he was told that the prawns would be irradiated with only 2 kGy rather than the customary 3.5-4 kGy as this would make the process undetectable.[15] Presumably, port health authorities would be confused because 2 kGy would reduce the bacterial load to a legal level but leave enough residual contamination so as not to arouse suspicions. Consumers, however, might not appreciate this practice, especially if the product were part of a previously rejected consignment.

Further evidence of 'Dutching' is provided by another UK newspaper investigation. In August 1989, the *Sunday Times* revealed how a 16-ton consignment of Indian prawns, condemned by US authorities because of *Salmonella* contamination, was being offered for sale by Landauer, a London firm of commodity brokers. One of Landauer's dealers is reported to have offered to send the prawns 'on holiday to Holland' — another trade expression for irradiating the food in the Netherlands. Rinus Verwijs, a Landauer company director, said,

> We tell our customers it has been rejected. What they do with it is entirely up to them. We would place them through irradiation for customers.

A potential profit of £100,000 after irradiation is a considerable incentive for a company to take up the offer, provided they can evade controls on such abuse[16] (see Appendix B4).

Can it be controlled?

Port health officers and honest food traders have told us that they believe the documented cases of abuse are just the tip of the iceberg. In the absence of detection methods, it is difficult to see how food law officers can prevent this abuse of irradiation by some sections of the food trade. This issue of controls is perhaps the most serious unresolved question over irradiated food. It is the basis for the demand, made by many organizations, that irradiation of food should not be permitted at least until there is a test to detect its use. This is a concern that has united groups on both sides of the debate about irradiation. It worries many honest traders who see the practice damaging the image of their product. It concerns large food manufacturers and supermarkets who could, quite unwittingly, be caught up in another major scandal. Above all, it concerns those people intimately involved in enforcement on whom the public rely to ensure food quality standards are maintained.

In the UK, local government environmental health and trading standards officers enforce the food laws. In many other European countries the controls are provided by national government. In North America some are federal and some are state or (in Canada) provincial functions. Most countries provide some means of enforcing food regulations which are designed to protect the consumer.

Whether one regards the process as beneficial or not, consumers need protection from fraudulent practices and misrepresentation of irradiated food. One of the strongest arguments expressed by some conservative Members of the European Parliament in the 1987 European Parliament debate on irradiation was that, without methods for detection, enforcement of labelling laws was impossible. The MEPs said they were simply not prepared to pass laws that were unenforceable.[17]

There is also a need to protect the consumer from possible health risks. Irradiated food may appear to be wholesome but contain hidden microbiological hazards. The important thing to note is that the abuse of irradiation we have uncovered is not merely deceitful, fraudulent or illegal. It could be a public health hazard. The problem is that irradiation has made obsolete the current methods for determining whether a food is safe, wholesome and fit to eat.

There are, essentially, three tools available. These are one's eyes, one's nose and a 'bug count', as the laboratory technique for measuring the bacterial load is known in the trade.

- A visual inspection will only spot the more obvious defects, and, with irradiated food, our eyes can mislead — the food stays looking fresh for longer.
- Smelling food will detect 'off' odours but irradiation can destroy the bacteria that give a warning smell.
- Testing for the general level of bacteria indicates whether the food is badly contaminated but irradiation kills many of the common bacteria leaving the food microbiologically 'clean'.

Killing bacteria may sound like good news. It is certainly promoted as one of the major benefits of irradiation. However, in the real world of food law enforcement, the bacterial count is used as an indicator test. A high bug count indicates a need for further testing for:

- *specific food-poisoning bacteria*. Irradiation will not kill all bacteria. Some of the more dangerous ones can be left behind but, now, in an environment where there are no competing micro-organisms.
- *viruses*. Food with a high bacterial load may have been harvested or processed in unhygienic conditions and may also be contaminated with viruses such as hepatitis. Irradiation will not kill viruses. These are much harder to detect and therefore can remain hidden.
- *bacterial toxins*. Irradiation will not remove the toxins, the chemical poisons created by some bacteria during the earlier stage of contamination before irradiation. With some

food poisoning bacteria it is these toxins that are the real public health hazard.

Thus there are hidden hazards which routine detection methods do not reveal.

The view of the UK government, the European Commission and the UN agencies involved in promoting food irradiation is that detection methods are useful but not essential. Control of irradiation is to be based on licensing and inspection of irradiation plants and inspection of documents which accompany irradiated food. It does not need very much imagination to think of a variety of ways that such 'paper' controls could be evaded. Some food traders already attempt to evade customs duty by re-invoicing to conceal either country of origin or true value, as the *Sunday Times* investigation of the contaminated Indian prawns revealed.[18] For all its inspections and controls, the US Department of Agriculture detected only one illegal shipment of irradiated pork, on its way to Sweden, when the ship was in mid-Atlantic.[19] How many others will go undetected?

The answer to the problem of food poisoning?

To the promoters of food irradiation, the issues of detection and control of irradiation are not very important. Irradiation, they argue, kills bacteria and makes the food safer. It is an essential tool in the fight against food poisoning. This is a highly simplistic view of the microbiological safety of irradiated food.

There is no dispute that there is a serious problem with the rise in food poisoning outbreaks. In the UK alone there has been a three-fold increase in reported cases since 1983.[20] Unreported cases are estimated at between ten and one hundred times the number of reported cases. Many people in the UK became concerned about this in 1988 after publicity over outbreaks of *Salmonella* and botulism food poisoning and the discovery of *Listeria* in a wide range of processed, chilled and ready-to-eat foods.

Making the problem worse

Is irradiation part of the solution to these problems? The short answer is 'no'. Our investigations suggest that irradiation may actually make them worse.[21]

Not all bacteria are harmful. Some are actually beneficial. Normally, food has a delicate balance of micro-organisms in competition. Eventually, their numbers increase and food which is not eaten slowly spoils. Put simply, good food goes bad. Unfortunately, irradiation does not kill all the different types of bacteria equally. Some are more resistant than others. By selectively reducing the numbers of some, irradiation alters the delicate balance of competing micro-organisms. In many parts of the world, microbiological control of food spoilage is achieved by maintaining this balance. A microbiologically sterile food is one in which any bacteria that are brought into contact with it will grow very rapidly.

Botulism

For example, irradiation can be used to reduce the level of *Salmonella* contamination on chicken or seafood. In the process it will also destroy yeasts, moulds and other bacteria, including those that give a warning smell when the food is going 'off'. Some bacteria, however, are not killed by irradiation and these will grow more strongly as the competition from other micro-organisms has been removed. This can be a serious hazard if spores of *Clostridium botulinum* are present and if the food is stored in a warm place and air is excluded — either in the centre of a large food item or by sealed packaging.[22 23 24] Thus, irradiation to reduce the risk from one food-poisoning organism can actually increase the risk from another. *Salmonella* usually only makes people ill for several days; botulism often kills.

The problem of not having enough bacteria to give a warning smell is disputed in the case of irradiation of fish where lower doses will be used. It is claimed that the food will smell unacceptable when the botulism becomes a hazard.[25] On the other hand, there is evidence that different strains of *Clostridium botulinum* and other bacteria vary widely in their sensitivity to irradiation. It has also been found that, while storage times can be lengthened with increasing doses of irradiation, the time taken for surviving *C. botulinum* to produce dangerous levels of toxins becomes shorter.[26 27]

Recontamination

Such possibilities reinforce the need for strict control of both the irradiation process and the conditions under which irradiated food is packaged, stored, and handled. This may severely limit the uses of irradiation. The US Department of Agriculture has effectively banned vacuum-packaged meat products from being irradiated.[28] It is clear from our conversations with USDA staff that botulism is the main concern. If irradiated foods cannot be packaged so as to prevent re-contamination by bacteria, the technology may have very limited use in the fight against food-borne disease.

Irradiation may create a false sense of security in the processing and handling of food. Total elimination of all bacteria cannot be guaranteed. Regrowth of remaining bacteria is always a possibility. Food can also become re-contaminated through contact with kitchen surfaces, utensils or a person's hands. Irradiated food needs as much if not more care in subsequent handling than unirradiated food if unforeseen microbiological problems are not to arise. This means that even more rigorous hygiene will be needed along with chemical preservatives and refrigeration to slow the growth of the remaining bacteria.

Mutations

Irradiation can cause mutations in bacteria and viruses in food leading to more resistant strains.[29] Strains of resistant *Salmonella* have been developed by repeated irradiation under laboratory conditions. Radiation-resistant bacteria have been found in environments with high natural or artificial radiation levels[30] and development of such resistance might be a problem around large irradiation plants.[31] As the UK Advisory Committee states,

> sub-lethal doses of ionizing radiation can produce chemical changes in genetic material of micro-organisms (mutations) leading to altered characteristics which will be propagated in subsequent generations. Such mutant micro-organisms could be more pathogenic than native forms. Also they might exhibit altered growth characteristics which would make them difficult to detect or identify, and thus interfere with the standard

> microbiological evaluation of irradiation. Mutants might also be more radiation resistant, and if they were to spread into the environment, they might contaminate food prior to irradiation and so render the process ineffective.[32]

Some new radiation-resistant bacteria have been detected on irradiated pork. Numbers of the bacteria present are small and the strain is not thought to be harmful but the experimenters conclude that,

> if low-dose irradiation were to be used to extend the shelf-life of fresh meats this organism, and possibly others, would not be killed.[33]

The problems of increased radiation resistance, increased pathogenicity and stimulation of growth of some surviving bacteria were recognized by the UN Codex Alimentarius Commission's Food Hygiene Committee in 1979. It is clear from the report of the 1982 meeting of the Codex Committee that there was a vigorous debate and disagreement among the various experts present. The meeting dismissed the problems but the points made by the Food Hygiene Committee are still valid, namely:

- the problems are difficult to investigate because no one has studied all surviving bacteria on irradiated foods
- irradiation does not remove all bacteria and can make some more resistant and possibly more dangerous
- more extensive knowledge will only come about when irradiation is more widely used.

Perhaps the strongest point was made by Dr Robert Charles of the UK Department of Health, who said,

> The consumer might well ask why it was then necessary to irradiate the food since irradiation did not provide any safeguards that could not be provided by thorough cooking . . . Furthermore, if there was any health risk due to induction of genetic mutations [these risks] would be added to those of other unavoidable processes such as cooking.[34]

Staphylococcus and *Listeria*

Irradiation will kill some bacteria but it will not remove the toxins that these bacteria have already produced. With some forms of food poisoning it is the toxin not the bacteria that represents the real health risk. For example, *Staphylococcus*, commonly transmitted to food by unhygienic handling, produces a toxin which is not destroyed by irradiation.[35] The same concern applies to *Listeria* which has been found to contaminate some milk products and many ready-to-eat, cooked-chilled foods. *Listeria* is a serious problem as the bacteria survive some heat treatments and grow at normal refrigeration temperatures. Use of irradiation to kill *Listeria* will therefore need to be 100 per cent effective — something almost impossible to guarantee. The *Listeria* toxin is of particular concern during pregnancy and there is some evidence that subsequent pregnancies may be affected long after recovery from the initial bout of listeriosis. The toxin is also known to be more virulent when stressed by heat or cold. Even though the UN agencies promoting food irradiation have claimed that it can help deal with the problem of *Listeria*, we have been unable to find any research on the effects of irradiation on the virulence of the toxin produced by the surviving bacteria. In the absence of such research, we believe these claims to be irresponsible.

Aflatoxins

Another example of irradiation actually stimulating production of poisons in food is found in the case of aflatoxins, potent liver cancer causing agents which have also been linked to suppression of the immune system.[36] Aflatoxins are produced by strains of the mould *Aspergillus* which is found on many foods grown, harvested or stored in warm, damp conditions. Like bacterial toxins, aflatoxins are neither removed nor reduced by irradiation. There is evidence that irradiation of vegetables and grains, at the doses being considered for commercial use, may *stimulate* aflatoxin production.[37] Table 7 indicates some of the increases observed. There can be wide variation in aflatoxin stimulation depending on the food, the strain of the *Aspergillus* mould and the doses used. Studies in the USA[38] found maximum aflatoxin levels on irradiated rice 50 times greater than on unirradiated rice. These studies also found that the increase in growth was greater if the *Aspergillus* was introduced to the rice after irradiation — suggesting that the problem may be greatest where the food becomes recontaminated after irradiation.

Table 7: Stimulation of aflatoxin production after irradiation[39]

Food	Percentage increase
wheat	increases with dose
corn	31
sorghum	81
millet	66
potatoes	74
onions	84

This is another highly controversial safety issue. The international Expert Committee dismisses the evidence indicating a cause for concern. It argues that any increase in aflatoxin production will be offset by reduced numbers of the *Aspergillus* mould. Other researchers have suggested that, as with the controversy over toxicological effects discussed in Chapter 3, the UN Agencies are taking too simplistic a view and that there has been selective, biased and misleading use of the available scientific evidence.[40]

A review of the evidence, commissioned by the Minister of Health in Australia in 1987, recommended that, if allowed, a minimum of 6 kGy should be used on irradiated grains and peanuts.[41] Unfortunately this would be well above the level at which unacceptable effects on flavour will become apparent.[42] Irradiation may be inappropriate for these foods unless other measures are taken to control *Aspergillus* mould and aflatoxin production. Humidity control during storage becomes even more important in the case of irradiated than non-irradiated foods.

Aflatoxins can be a special concern with peanuts. Importers and processors, particularly in the peanut butter trade, monitor aflatoxin levels on imported nuts. They fear that, if irradiation is legalized, suppliers may irradiate nuts to reduce *Aspergillus* mould levels. As a result, competitor micro-organisms will also be destroyed. After the importers have purchased what they believe to be 'clean' peanuts, aflatoxin levels could increase rapidly, either by re-growth of remaining mould or from re-contamination. It is the peanut processors not the foreign suppliers that the consumer will blame for any subsequent risk to health.[43]

Salmonella

Two-thirds of food poisoning outbreaks in the UK are linked to *Salmonella* poisoning from chicken, though beef and other foods may also be implicated. Many cases of food poisoning are linked to inadequate cooking and refrigeration and there is a tendency to blame the victims for their own misfortune. Certainly inadequate cooking at home is the cause in some cases but many more occur in restaurants and catering. Can the consumer really be blamed when 60-80 per cent of frozen poultry is contaminated with *Salmonella*?[44] No one goes shopping for *Salmonella*. Is the food industry saying that the public can no longer expect food free from contamination when it is purchased? We believe that the public has a right to expect clean food and food industry should provide this — by regulation if necessary.

It has been argued that the problem of *Salmonella* contamination in poultry has now reached the stage where irradiation is necessary — other methods on their own being insufficient to deal with it. It has even been argued that irradiation of all poultry should be compulsory rather like pasteurization of milk.[45] This position is defeatist, damaging to the poultry industry and wrong. Some poultry processors in the UK claim to have been able to eliminate *Salmonella* contamination by detailed attention to hygiene.[46] In 1989, the World Health Organization reported on advances in oral vaccines capable of producing *Salmonella*-free flocks.[47]

It will certainly not help chicken sales if producers have to admit to using irradiation. Quite apart from the resistance to irradiation itself, its use would be an admission of a breakdown in hygiene during processing. The principal source of *Salmonella* in chickens is contaminated feed. The problem of feed contamination is often made worse by recycling of chicken carcasses, some of which may be contaminated, back into chicken feed. It is not the meat itself but the bowel of the chicken which is contaminated with *Salmonella*. Let the chicken industry tell consumers that current methods for processing chicken have burst the gut, spread faeces over the carcass, failed to clean this faecal matter off thoroughly and then irradiated the poultry meat to hide the resulting *Salmonella* contamination. See what that does to chicken sales!

Major UK chicken producers have already declared they have no interest in irradiation. They say they cannot market irradiated chicken on the basis of reducing *Salmonella* contamination.[48] They claim irradiation produces unacceptable colour, flavour and texture changes in poultry meat and adds significantly to the price.[49] The UK poultry industry suffered a 30 per cent reduction in egg sales in late 1988 as a result of a statement by a Junior Health Minister that egg production was contaminated with *Salmonella*.[50] It clearly has no wish to repeat the lesson with chicken sales.

In any case, irradiation is not particularly cost-effective. A Canadian study[51] lists benefit/cost comparisons for 11 *Salmonella* control options. Public education, use of chlorinated chill water in packing houses, adequate cleaning and disinfection of poultry crates are all seen as 5-8 times more effective than irradiation. Other disinfection and feed sterilization techniques are as effective as irradiation. The World Health Organization, however, claims that there are considerable benefits from irradiating poultry.[52] This claim relies on the assumption that all cases of *Salmonella* food poisoning can be eliminated by this method. This would imply that most, if not all, poultry would have to be irradiated. The findings of a Scottish study[53] suggest that irradiation would only be economically cost-effective where there is a continuing high level of food poisoning — the implication being that it is justified only if other effective measures to reduce *Salmonella* contamination are not implemented. If any of these other measures are used, irradiation might not be an economically justifiable proposition.

As with other bacteria, different strains of *Salmonella* exhibit widely differing sensitivities to irradiation. Several countries permit irradiation of chicken for salmonellosis control up to 7 kGy,[54] but this dose may not be sufficient to eliminate all strains. In the real world, odours and flavour changes are detectable at doses above 3 kGy. Thus, elimination of *Salmonella* by irradiation may not be feasible. The temptation will always be to use lower doses to minimize the effects of irradiation on taste and smell — thereby leaving a level of residual contamination and a false sense of security if it is marketed as '*Salmonella*-free'.

Food hygiene or food irradiation?

Even if it were effective, irradiation to deal with food poisoning organisms is not something which ought to be encouraged. Microbiological contamination of food with harmful bacteria is, in all cases, a result of a breakdown in hygiene in the processing and handling of food. There is much that needs to be done to improve hygiene — not just in developing countries. The UK Ministry of Agriculture's meat hygiene division found that of 919 slaughterhouses in the UK, the standards in 845 (90 per cent) were so low that the meat could not be sold for export to the European Community.[55]

The promoters of irradiation appear to agree that it should not be used as a substitute for hygiene controls — even though the cases of abuse we have uncovered show that it is being used in this way already. The statement from the 1988 International Conference on the Acceptance, Control of and Trade in Irradiated Food states,

> Food intended for treatment by irradiation should be of a quality acceptable for Good Manufacturing Practices (GMP). Hygienic practices which are needed in GMP for other processes are also necessary in the process of irradiation, but irradiation should not be used as a substitute for such practices.[56]

The problem is that, until 1989, these same UN agencies consistently refused to accept the need for international standards that would define when 'hygienic' and 'good manufacturing' practices have not been used. Only now are the agencies discussing microbiological standards for foods prior to irradiation. However, the standards proposed for seafoods are clearly designed to permit irradiation to continue to be used on food of dubious hygienic quality. There is disagreement even among the

UN agencies over this. The UN Food and Agriculture Organization has supported programmes which have been effective in improving hygiene in seafood processing in India. The FAO has criticized the proposals of the International Atomic Energy Agency that would permit use of irradiation rather than hygiene to achieve acceptable microbiological standards for products of some South-east Asian countries. It argues that this will forever condemn these products to being seen as low-grade and second-rate in the international market.[57]

The use of irradiation is a clear warning sign that all is not well with some foods. Some sections of the food trade are aware of the dangers that irradiation poses to the image of their product. The image of shrimp and prawns has already been damaged as a result of the scandals that have come to light. These scandals have hurt the reputable traders as well as the unscrupulous. It has hurt the image of the product from India and South-east Asia regardless of whether the problem originated within these countries or occurred as a result of poor handling within the international trade.

There are companies that have never needed irradiation and who have declared that they will not use it. We applaud the courage of those which, like Ken Bell International, have decided to 'blow the whistle' on the abuse of irradiation within the seafood trade.[58] The problem is not with seafoods or with any other group of foods, but with the unhygienic practices that lead to some of these foods being contaminated.

The food industry faces a clear choice — improve hygiene in the processing of food or use irradiation to hide the contamination. The consumer faces a similar clear choice — to buy clean food, thereby encouraging the companies that pay attention to hygiene, or to accept irradiated food and legitimize the worst hygiene practices.

Is irradiation safer than other methods of preservation?

Irradiation is promoted as an alternative to some chemical methods of preserving food. This sounds useful as some chemical additives and pesticides are now being banned by many countries.[59] It has been suggested as an alternative for preserving fruit, grains, spices, potatoes, onions and garlic. How realistic are these claims?

Fruit

Of 27 fruits investigated, only 8 have been found appropriate for irradiation (see Table 8). Irradiation causes considerable damage to fruit tissues at even small doses.

Table 8: Response of 27 fruits and vegetables to irradiation[60]

	Beneficial effects			Not beneficial		
Fruit	(a)	(b)	(c)	(d)	(e)	(f)
Bananas	*					
Mangoes	*					
Papayas	*					
Sweet cherries		*				
Apricots		*				
Tomatoes		*				
Strawberries			*			
Figs			*			
Pears			*			
Avocados/nectarines				*		
Lemons				*		
Peaches				*		
Grapefruit/pineapples					*	
Oranges/lychees						*
Tangerines/honeydew melon						*
Cucumbers						*
Summer squash						*
Bell peppers						*
Olives						*
Plums						*
Apples						*
Table grapes						*
Cantaloupes						*

(a) delay in ripening (b) delay in ageing
(c) control of storage decay (d) damaged by radiation
(e) accelerated ripening (f) no positive benefit

Even where it is considered 'appropriate', irradiation of fruit is often unable to provide any real extension of shelf-life. Much fruit for export is currently picked slightly under-ripe and allowed to ripen in transit. If it is irradiated before ripening has begun, it frequently never ripens. The extra one or two weeks" extension of life by irradiation is largely lost by the inability to ship under-ripe fruit. Ripe fruit bruises more easily. Irradiated fruit bruises easier still. In practice irradiation offers little advantage to fruit exporters.

One of the strongest arguments that can be made for irradiation of fruit is the need to control insect infestation, in particular to prevent insect pests migrating in fruit to areas where they are not currently a problem. Ethylene dibromide, which was commonly used to control the spread of fruit fly, is now banned in the USA and Europe. This presents a problem to growers in some areas where the fruit fly is endemic, who wish to export their produce. Is irradiation an alternative to the use of this chemical? Irradiation will kill insects but the dose needed usually results in unacceptable damage to the fruit. Irradiation at lower doses will 'sterilize' insect larvae — they are unable to make the transformation into adult flies or, if they do, these flies are unable to reproduce.

If irradiation is used for insect control, it will not replace the use of pesticides before harvest. These will still be needed. There is some concern over the effect of irradiation on pesticide residues in food — an area where the UK Advisory Committee acknowledges there has been too little research.[61] There is also some doubt about whether irradiation can be used as an alternative post-harvest treatment. Port health authorities have yet to agree that a piece of paper saying fruit has been irradiated is acceptable for quarantine purposes. Any inspector who observes moving insects or larvae may have good reason to suspect that not all are sterile. A minimum dose of 0.15 kGy is necessary to guarantee sterility of fruit fly larvae — i.e. that they would not produce fertile adult flies — but a dose of 0.35 kGy will cause unacceptable damage in tropical fruit such as mangoes.[62] In commercial use, the difference between maximum and minimum doses to a food batch can often be a factor of two to three. The process is therefore on the margin of feasibility. There is little or no margin for error.

In practice, if fruit damage is likely to occur at the upper end of the dose range, there will be strong commercial incentives to reduce the average dose. The result could be insufficient exposure to guarantee sterility of some of the larvae. A port inspection will have no independent means of verifying the dose applied to the fruit. If the irradiation plants do get it wrong the effects can be very damaging. The population of California can still recall how in 1981 Governor Gerry Brown turned a very small fruit fly problem into a much larger one by agreeing to the release of large numbers of supposedly sterile male insects that had been irradiated in Mexico. These were supposed to mate with fertile females who would then not produce viable offspring. Unfortunately, the irradiated males turned out to be less sterile than they should have been and the governor ended up having to order the bombing of populated areas with the pesticide malathion from helicopters at night. Governor Brown was defeated at the next election.

The argument that irradiation is essential was used in Hawaii when ethylene dibromide was banned in 1984, threatening Hawaiian papaya exports to Japan and mainland USA. However, Japan would not permit imports of irradiated papaya. Since irradiation was not economically viable for US exports alone, the exporters were stimulated to develop a steam heat treatment for the exports to Japan and a 'double-dip' liquid heat treatment process for export to the US mainland. These have proved effective. The only cases of problems with the double-dip method have been traced to produce exported by the papaya packer which had been promoting irradiation on the Islands. An investigation by the US Department of Agriculture suggested that the high level of infestation found in the fruit when it reached California was consistent with the possibility of deliberate contamination and this could not be ruled out.[63] An irradiation plant would barely be viable even if all the Islands' papaya were irradiated and this would mean abandoning the heat-treatment plants already built. This is an example of how pressure to find an alternative to both chemicals and irradiation has led to technological advance.

Irradiation has been suggested as a suitable quarantine treatment for the mango crop of many countries. It is being promoted by the IAEA for countries in the Asia/Pacific region. South Africa already irradiates mangoes commercially. In Queensland, Australia, irradiation was promoted as essential for mango exports. Unfortunately, the variety of mango grown in Queensland is softer and more easily damaged by irradiation than the South African variety. Queensland also has a problem with the mango seed weevil. Once ripening has begun, the doses needed to sterilize the weevil cause severe damage to the fruit. These differences in the suitability of irradiation for different varieties of some foods concerns consumers as well as growers. The varieties of mangoes, strawberries and other fruits which will withstand irradiation are harder, 'woodier' and generally lacking in flavour. Will widespread use of irradiation lead to a change in the varieties grown — with a predominance of those which can withstand irradiation damage? If the UK market is dominated with strawberries like those which are currently irradiated in France, is there any real improvement in either quality or choice?

Consumer reactions to irradiated fruit are claimed to be favourable. The IAEA claims that irradiated fruit outsold the conventionally treated counterpart in Thailand, the USA and France.[64] The facts are very different. The US trial for irradiated papaya had to be abandoned because of consumer opposition. In France the consumer rejection rate for irradiated strawberries was 60 per cent and only 25 per cent of the original purchasers returned to buy them a second time.[65]

Even if irradiation could be used for quarantine control of insects, consumers may not be pleased to find larvae or insects in their fruit even if these have been sterilized. Once again people may choose to ask, if the fruit had to be irradiated what was wrong with it?

Grains and spices

Food traders in many countries are seeking alternatives to ethylene oxide gas which is used to fumigate grains and spices. Might irradiation be a suitable substitute? The world's largest irradiation facility at Odessa in the USSR is used to disinfest grain imports. South-east Asian countries are interested in irradiation of rice exports mainly for the Japanese market.[66] Unfortunately, Japan does not permit irradiation of rice and even at doses as low as 0.5 kGy, there are noticeable changes in rice flavour.[67]

There is the unresolved question of storage times for irradiated wheat and possibly other grains, and the controversy over stimulation of aflatoxin production in grains. These ought to be resolved before the process is widely used. It is worth noting that the USA has permitted irradiation of wheat since

1963 but there has been no significant use for this purpose. Alternatives to either fumigants or irradiation include oxygenless air, cold storage, or heat treatment. Some stored grain in Australia is treated with carbon dioxide gas. Provided the grain silo is properly sealed, its contents can generally be stored for four months without infestation.[68]

The spice trade will be hard hit if more countries ban ethylene oxide. Could irradiation be used instead? This is one area where some critics of irradiation are prepared to make concessions.[69] Spices are, after all, eaten only in small quantities. However, alternatives do exist for many spices. A Danish company has developed a protein coating that will allow many spices to be heat treated. The process works well for pepper but not for spices such as paprika where the colour is important.

There is, however, a more important question to ask. How is it that spices which have been used for centuries as preservatives have become a potential source of food contamination? There is something very wrong with the spice trade. These problems are the result of a breakdown of hygiene in the production, processing, handling and storing of spices. If the water used in spice growing is contaminated and if spices are dried and stored in conditions where insects, rats and birds leave their droppings, then obviously they become contaminated. Consumers may prefer to buy clean spices rather than those contaminated with faecal matter and food-poisoning bacteria which have been concealed by irradiation. Put bluntly, filth is filth whether it is irradiated or not[70] and we, like many other people, would prefer not to eat it. Accepting irradiation for spices inevitably means there will be less incentive to tackle the root problems of bad hygiene.

Potatoes

Irradiation has been suggested as an alternative to chemical treatments which inhibit the sprouting of potatoes. Irradiation also inhibits the production of (harmless) chlorophyll that gives potatoes a green colour when they are exposed to light. This green colour often indicates the presence of a toxic alkaloid called solanin which is also produced by exposure to light. Solanin should be avoided during pregnancy as it carries an increased risk of spina bifida. Unfortunately, solanin production is not inhibited by irradiation. Consumers may no longer have the obvious colour indicator that potatoes may be harmful.[71] [72] [73] [74] In most western countries potatoes are sold loose or in clear plastic packs because consumers want to inspect them for cuts and mould. Irradiated potatoes, if they are permitted, will need to be sold in light-tight packaging.

Irradiation will increase bruising of potatoes and slow the healing of any cuts. For this reason a delay after harvesting is recommended before irradiation. Irradiation also increases the susceptibility to fungus attack and rotting. Careful inspection and sorting must remove soil and any rotten or damaged potatoes, as otherwise the whole batch will quickly spoil. These factors, brought on by particularly adverse damp harvesting conditions, are thought to be reasons for failure of a Canadian potato irradiation project in the mid 1960s.[75]

Denmark recently revoked the permit for potato irradiation which was granted in 1970 as it had never been used. Japan granted permits for irradiation of potatoes in 1972. Japanese authorities have withheld information on sales of irradiated potatoes since 1976 but they are believed to have fallen from a peak of 21,000 tons to around 7-8,000 tons in 1987. Technical reports on the Research and Development Programme of the Japan Atomic Energy Commission have also been witheld from the public. Leaked data indicate that these show rats fed irradiated food had lower growth rates, increased mortality and increased ovary weight.[76] Irradiation of potatoes is not likely in the United Kingdom.[77] The main concern would be imports from other European Community countries though even this seems unlikely. In theory irradiation could be used to inhibit potato sprouting; in practice it seems unlikely that it will.

Onions and garlic

Unlike potatoes, onions are best irradiated within four weeks of harvesting. Damaged or rotting onions also easily spoil the whole batch. Onions and garlic both show signs of internal browning as irradiation kills the growing tip. This lowers their commercial value. International Atomic Energy Agency studies in 1988 showed that irradiation required additional costs in the sorting of damaged onions prior to irradiation as well as the additional costs of irradiation. The use of irradiation was economically viable only in the fifth month after harvesting, there being no economic advantage in months one to four or in month six.[78] This was in a South-east Asian country where there were strong economic incentives to find a way of storing onions longer so as to reduce imports at various times of the year. It was also a country where labour is cheap and hand-sorting to remove rotten and damaged onions is feasible. It is unlikely that irradiation of onions or similar foodstuffs will be economically viable in Europe, North America or Australasia.

Japan has not given a clearance for irradiation of onions. Animal feeding trials in Japan are believed to have indicated adverse effects. The studies have not been published but leaked reports suggest abnormal organ weights and bone deformities.[79] It seems that there may be a case for witholding permits for irradiation of onions on safety grounds. This issue should be included in the review of safety which the World Health Organization is being asked to undertake.

Irradiation — an alternative to additives?

Another promotional argument for irradiation is that it will reduce the need for harmful chemical additives in food. There is a very real and well-founded public concern about the extent to which foods are being adulterated with chemicals. Processed foods rely on additives not merely for preservative effects but for flavours, colours, and bulk fillers too. There are good reasons for concern about the effect of pesticide residues[80] and food additives on both consumers and workers.[81]

Far from eliminating additive use, the process of irradiation will itself *require* use of a number of additives in order to control some of the undesirable effects.[82] Table 9 lists some of the additives that are claimed to be eliminated or reduced by irradiation *and* those proposed for use with irradiation. Note that some of these (marked *) are identical.

Additives will also prevent discoloration and bleeding in meat. Table 10 outlines a series of steps to control such effects.

Note that the extension of shelf-life is a gain for the processor and retailer. The consumer gains nothing — except the packaging and additive — from this process. A telling comment from within the retail trade came from one of the directors of Coles-Myer, one of Australia's leading retailers,

> If you have to transport fresh food to and from an irradiation plant it is going to need to have a longer shelf-life to make up for the extra time it takes to get it there and back again. We are not in the business of subsidizing the transport industry.[83]

Table 9: Irradiation and additives[84,85,86]

Some hazardous additives which it is claimed irradiation might replace

Additive		
* E251 Sodium Nitrite		m, c,
* E230 Diphenyl		c,
E220 Sulphur Dioxide		c,
E210 Benzoic Acid		a, i,
Propylene Glycol	c,	
925 Chlorine		i,
E926 Chlorine Dioxide		i,
E281 Sodium Propionate	a,	
E236 Formic Acid		i, (banned in UK)
Ethylene Oxide		c,i,m,t,
Methyl Bromide	i,	
Ethylene Dibromide		i, c,
Propylene Oxide	i,	
Hydrogen Peroxide		i,

Some additives which might be needed to reduce undesirable effects of radiation

Additive	
* E251 Sodium Nitrite	m, c,
* E230 Diphenyl	c,
E221 Sodium Sulphite	i,
E300 Ascorbic Acid	
E321 B.H.T.	c,
E320 B.H.A.	c,
371 Nicotinic Acid	i, a,
924 Potassium bromate	i,
Sodium tripolyphosphate (TPP)	i,
Sodium chloride (salt)	
Glutathione (for Vit. B1)	
Niacin (for Vit. B6)	
Sodium Ascorbate (for Vit. C)	

Key:
i = irritant, c = carcinogen (known or suspected cancer-causing agent in animals and/or humans), m = mutagen (capable of causing mutations), t = teratogen (capable of causing damage to developing foetus), a = capable of causing allergic reactions.

Table 10: Steps for irradiation of meat to extend shelf-life of retail cuts to 21 days[87]

Step 1	cut into portions
Step 2	dip in a dilute solution of sodium tripolyphosphate* (or other condensed phosphate)
Step 3	filmwrap
Step 4	vacuum-pack filmwrapped portions in a bulk container
Step 5	refrigerate at 0-5 °C
Step 6	irradiate with a dose of 1-2 kGy
Step 7	ship and store at 0-5 °C
Step 8	remove from container for display (no more than half an hour before display)
Step 9	display refrigerated at 0-5 °C
Step 10	sell within 3 days[88]

* Sodium tripolyphosphate is a chemical used for cleaning grime off walls (it cuts grease). It is also irritating to the skin and is used as a purgative. Its use here in conjunction with irradiation is to bind water into the meat — why sell meat when you can sell water?[88 90]

In the UK, Marks & Spencer, whose quality standards for food are recognized to be among the highest, has said that irradiation does not fit in with the company policy of providing good quality fresh food on a fast turnover.

Irradiation destroys some enzymes and increases the activity of others. It is proposed to treat slaughter animals with adrenaline if the meat is to be irradiated.[91]

Some of the additives in Table 9 are vitamin supplements designed to make good the losses discussed in Chapter 4. Some are antioxidants presumably included to deal with the peroxidation/free radical chain reactions discussed in Chapter 3. To this list might also be added vitamin K, vitamin E and vitamin A (see Chapter 3). Perhaps, in view of the concerns raised over such losses, there is a case for requiring vitamin supplementation of some irradiated foods. Even so, there is evidence that added vitamins are less effective than those occurring naturally.[92] There will also be increased need for artificial colours and flavours to mask some of the 'off' flavours and colour changes produced by irradiation.

It appears that, like many of the other promises, irradiation is unlikely to be a usable alternative to chemical additives. In fact, further adulteration of food may be needed in order to make irradiated food look and taste acceptable.

Worker and environmental safety

In Chapter 2 we outlined how an irradiation plant is supposed to work. What actually happens? What are the risks to workers in irradiation plants, and people who live near such facilities?

Advocates of irradiation lay great emphasis on the fact that such facilities will operate under internationally agreed standards for radiation protection.[93] They conveniently ignore the fact that these standards have been challenged by environmental groups, some independent scientists, and many trade unions.[94]

The simple fact is that there is no safe level of radiation — any exposure can be the one that does the damage that may show up as cancer or genetic damage in future generations. Protection standards are a balance between these risks and assumed benefits — the so-called 'acceptable risk' philosophy. The protection agencies can be shown to have the balance of risks and benefits very wrong. The risk estimates recommended by the International Commission on Radiological Protection have been shown to be at least 2-3 times (some suggest 5-10 times) too low.[95] [96] The permitted dose limits for workers and the public are too high by the same factors. Despite this, governments have been revising regulations for radiation protection in line with the latest international recommendations.[97] These actually relax the current standards in some critical respects. For example, doses to internal organs were permitted to increase by 2-10 times the old levels. Many of the limits on intake of critical radioactive materials were similarly relaxed.[98]

Irradiation is an extremely hazardous process. Exposure to the unshielded source can deliver a lethal dose. Even routine exposures to much lower levels could lead to long-term effects.

The radioactive materials have to be transported into the plant, the spent sources (still radioactive) removed and any contaminated material disposed of. At each stage in the process there is the possibility of accidents leading to exposure of workers, the public and the environment.

Accidents at irradiation plants

These are not hypothetical concerns as the history of accidents and violations of worker and environmental safety in irradiation facilities clearly shows. Among those which have come to light are the following.

Radiation Technology, New Jersey, USA

In 1986, the US Nuclear Regulatory Commission revoked the licence of one of Radiation Technology Inc's irradiation plants for a series of wilful violations of worker and environmental safety, of which the NRC stated,

> [the violations were] wilful and numerous management operations personnel wilfully provided false information to the NRC thus demonstrating a pattern of wrongdoing so pervasive that the NRC no longer has reasonable assurances . . . that the licencee will comply with NRC requirements and that public health and safety will be protected.[99]

In all, the company was cited 32 times for various violations including:

- bypassing of safety interlock systems during which one worker received a dose of radiation in excess of permitted limits in 1977
- an accident involving a leaking cobalt 60 source in 1975 and illegally burying radioactive material on site.

Martin Welt, the company founder who had ordered employees to lie to the NRC investigators, was forced to stand down as company president under pressure from the NRC. The US Department of Energy promptly hired him as a consultant on $100 an hour plus expenses to advise on the

DOE's 'By-products Utilization' programme which aimed to build six demonstration food irradiators in various states in the US. Welt has also advised the Queensland government in Australia on plans for construction of an irradiation plant in Brisbane. He was eventually convicted of conspiracy in the US courts in 1988.[100] It was revealed in court that the company had taken secret cash payments to irradiate food brought to the plant at weekends.[101] In September 1988 Welt was fined $50,000 and sentenced to two years in prison. He and a close relative, Andrew Welt, are directors of another company, Alpha Omega Technology, which has been seeking funds from the World Bank to build irradiation facilities in Malaysia and other South-east Asian countries. The company has also assured potential customers for irradiation in Hawaii that a plant will be built there by November 1990.[102]

Isomedix, New Jersey, USA

The NRC has cited Isomedix Inc, the largest irradiation company in the USA, for allegedly:

- overexposing workers
- failing to signpost radiation areas
- allowing food and cigarettes in radiation areas
- operating the system without authorized personnel being present
- failing to monitor water disposed to the sewer.

'Whistleblowing' by workers at one Isomedix plant led to the discovery that radioactive water had been disposed of down the toilets and contaminated the pipe leading to the sewers. The incident, which occurred in 1974, resulted in worker over-exposures and the radioactivity was still detectable in 1979.[103]

International Nutronics, New Jersey, USA

A nine-count federal indictment, issued in 1986, charged the company with the cover-up of a radioactive spill in 1982. Management ordered employees to dispose of the radioactive water down shower stalls, thus contaminating the sewer system. Workers cleaning up the spill were instructed to move radiation badges from belt to collar so that they recorded a lower dose. The contamination was still detectable outside the building 10 months later. The company was fined.[104]

Demonstration irradiator, Hawaii

A radioactive leak at an irradiator in Hawaii in 1967 contaminated the pond water, shipping cask, roof, machine room, tools and workers' clothing.[104a] The clean-up cost Hawaii $385,000 in 1979. In 1980 residual contamination at dangerously high levels was still detectable on the lawn outside the building.

Radiation Sterilizers, Georgia, USA

A leaking caesium source in June 1988 contaminated 25,000 gallons of pond water, workers' clothing, cars, and home carpets. It also resulted in contamination of some of the irradiated products: medical supplies, milk cartons, and saline eye solution. The incident was still, in the words of the investigating team, 'on-going' in 1989 with costs exceeding several million dollars. The team's report reveals that regulations governing commercial irradiation companies in the USA have not required effective radiation safety programmes and says,

> Now that such a high-consequence event has occurred, a new regulatory focus on health physics [radiation safety] programs for all irradiators should be considered.

The report also states,

> The RSI incident demonstrates the need for a detailed emergency plan . . . before a licence is issued.[105]

University of Tennessee, USA

As well as the overexposures of workers at Radiation Technology and Isomedix, there is the earlier case of overexposure, caused by a worker bypassing a safety 'interlock' in a dry storage facility at the University of Tennessee in 1971. The worker received a high dose but survived.[106]

Stimos, Pontecico, Italy

A maintenance worker entered the irradiation cell of a corn irradiation facility on the conveyor belt while the source was operational. He received a high radiation dose and died 12 days later.[107]

Institute for Energy Technology Irradiation Plant, Kjellet, Norway

Failure of safety devices allowed a maintenance technician to enter the irradiation cell with the source exposed. The technician died 13 days later.[108]

Juarez, Mexico

While not strictly an irradiation incident, the theft and melting down of a cobalt 60 source in Mexico[109] illustrates how easily widespread radioactive contamination can occur. The incident was detected when a truck carrying metal goods which contained some of the radioactive cobalt took a wrong turning in the USA and set off the monitors at a nuclear facility. Radiation was detected in a wide range of metal products, the highest levels being found in table legs in a Chicago restaurant.

Goiania, Brazil

A similar theft and dispersal of radioactive caesium 137 from an abandoned therapy unit occurred in Brazil.[110] Material was taken home in scrap-yard workers' pockets. One spread it on the floor of a child's bedroom because it glowed in the dark, another child sprinkled it on her arms. The incident resulted in thousands being monitored, some 100 people being found seriously contaminated, and several being hospitalized. Large areas of the city were found to be contaminated. Caesium is water-soluble, easily dispersed and almost impossible to recover.

Delmed, San Salvador, El Salvador

Three workers at the Delmed plant suffered serious radiation injuries when a Canadian-made irradiation equipment for sterilizing medical supplies malfunctioned in February 1989. The three are believed to have entered the irradiation cell without monitoring equipment and been exposed to one or more cobalt rods which had 'stuck above the water'. They received estimated doses in the range of 3-8 Gy and were hospitalized in Mexico City. On 10 February Delmed asked for help from the Canadian company, Nordion, which had supplied the plant. Nordion staff found the source rack to be unsafe and removed the cobalt from the irradiation cell. The Canadian company was not however told of any overexposure until 26 February.[111]

Steritech, Dandenong, Australia
Lack of local expertise to deal with emergencies in irradiation plants is not confined to 'Third World' countries. An inspection of the Steritech plant's records in 1986 by Australian member of parliament, John Scott, revealed that the plant had on one occasion been shut down for five days because a wire cable controlling the cobalt 60 source rods had jammed. This had prevented the rods being lowered into the pool. Appearing before the Australian Parliamentary Inquiry on Food Irradiation in 1987, George West, the Managing Director, revealed that a maintenance team had to be flown from Canada to remedy the problem. Technicians and equipment to deal with such emergencies are not available in Australia.[112]

Becton Dickinson, North Canaan, Connecticut
Aluminium boxes, containing products being irradiated, jammed into the source at a Canadian-designed irradiator in Connecticut in 1981. Operators were unable to lower the source. Despite this, the control panel indicated that the cobalt source had been safely lowered into its storage pool. It was only the monitors indicating high radiation levels that warned workers all was not well. Technicians had to be called in from Atomic Energy of Canada Ltd (as Nordion was then named). They managed to lower the source but dislodged several cobalt rods from the rack in doing so. These had to be recovered using long-handled tools and mirrors.[113]

Safety in the future?

What assurances are there that the regulations for worker safety, the planning, expertise and equipment for dealing with emergencies, and the operation of future plants being built all over the world will be better controlled? The above incidents all happened in plants built and operated under the internationally agreed system of regulation and control. The international promotion literature claims that,

> As far as workers in the food industry are concerned, irradiation poses no greater risk than other technologies for food processing. Indeed, irradiation is safer than

some food processing methods, such as those using hazardous substances.[114]

Trade unions with members in the food industry have taken a different view. Food industry unions have adopted a highly critical stance on this technology in the UK, throughout Europe, the USA, Australia and world-wide through the International Union of Foodworkers.[115] The unions give two reasons for their position. The first, they are rightly concerned about the safety of their members who might have to work in irradiation plants. The second, as summed up in 1985 by Nigel Bryson of the UK Bakers Food and Allied Workers Union, and a founding member of our working group on food irradiation, 'Our members are consumers too!'

A technology looking for a use?

It is hard to escape the conclusion that there are few if any benefits from irradiation. In the words of Dr Edward Radford, former chairman of the US National Academy of Sciences Committee on the Biological Effects of Ionizing Radiation, irradiation is a technology looking for a use. There are a number of scientific questions that have not yet been answered about the safety of the process. There are also questions about the manner in which the various national and international expert committees have reached their conclusions that it is safe. There are questions about the quality of irradiated food, the manner in which it will be used and the way irradiation is already being abused. It seems entirely reasonable to require that the strictest regulations, and systems for enforcement of these regulations, be in place before its use is permitted.

We say this as critics who after five years of asking questions have become less satisfied with the answers than we were at the beginning. From a position where we were prepared to concede that there might be some benefits from irradiation we now see none. We find that the use of irradiation should be a warning sign for consumers. What was wrong? Why was irradiation needed for the food product? What has been hidden? Good food did not need irradiation. If any food needed irradiation consumers will need to ask why. Who wants irradiation so badly that they are prepared to override these concerns?

6. Who wants it?

Late in 1986, one of our colleagues in Australia received an unlabelled computer disk in the mail. On it was the draft of a global marketing strategy for food irradiation devised by a group of public relations and marketing experts.[1] After the draft document was circulated world-wide among consumer organizations, a sanitized version was published by the International Atomic Energy Agency in 1987.[2] These documents acknowledge the problems that promoters of food irradiation are having in persuading governments, the food industry and consumers to accept that irradiation is either safe or useful.

Despite the approval given by the IAEA/FAO/WHO Joint Expert Committee in 1981, irradiation technology had clearly failed to 'sell itself'. A global propaganda initiative was needed to win the battle for the hearts, minds and money of consumers. In this chapter we will investigate the nature of this campaign, and who is behind it.

Promotion of food irradiation — the marketing strategy

The marketing strategy identifies the arguments and initiatives necessary to 'sell' irradiation to a sceptical public. The objective is to achieve world-wide acceptance of irradiation of food as safe, beneficial, and in some respects, necessary. It selects specific strategies and approaches to persuade:

- governments to permit irradiation and sale of irradiated food

- the food industry to adopt irradiation as a processing technique
- non-governmental organizations to endorse the views of the international promotion campaign
- consumers to buy irradiated food.

Different arguments are emphasized with each of the four main target groups. There is also a shift in emphasis among the benefits claimed for irradiation. The argument that irradiation will 'extend shelf-life' of food is better used with the food industry than with consumers who are unlikely to see this as a benefit. Irradiation will not be extending the time in the refrigerator or cupboard at home. It will extend the time in the supermarket, warehouse or in transport. Irradiation means older food to the consumer at a time when the market-place is responding to consumer demand for fresher produce.

The draft marketing strategy is quite clear about who stands to benefit from irradiation:

> The initial marketing of irradiation is not primarily aimed at consumers . . . Consumers will not ask for food irradiation. They do not feel the need for it, since they are not sufficiently aware of many of the present problems with food and the benefits the process offers.[3]

On the other hand:

> The food industry stands to benefit from virtually all the potential applications from food irradiation. The primary target of the food irradiation marketing plan . . . should be the food industry.[4]

Irradiation is to be sold to consumers by emphasizing two highly emotive concerns. These are:

- food poisoning — which is to be presented as a major (if not the world's number-one) public health hazard. Irradiation is to be offered as a solution by emphasizing its ability to kill bacteria
- world hunger — which is made worse by food losses. Preservation of food and control of insect infestation with irradiation are again offered as the solution.

In 1983, the International Atomic Energy Agency and UN Food and Agriculture Organization established the International Consultative Group on Food Irradiation (ICGFI). This body, supported by 26 countries (see Appendix C3) has been the major vehicle for promotion of food irradiation. The strategy urges the ICGFI representatives from the participating countries to:

- identify key decision-makers in government and influential non-governmental organizations such as medical, dietary and public health, environmental and consumer associations 'organizationally and personally'; contact these 'key decision-makers on a personal basis to gain approval of food irradiation'; and provide these decision-makers with 'complete information packages about food irradiation; its history, its safety, its efficacy, and consumer benefits'.
- set up 'lead organizations' in each country to promote food irradiation.
- approach 'food industry, irradiation processors, etc., to raise funds and promote food irradiation', work with industry to achieve a 'breakthrough with one or two manufacturers to "showcase" the acceptance of irradiation' — help these 'technically or in funding to achieve a commercial breakthrough' and develop a 'unified, professional public relations campaign' with a 'common message for retailers and consumers.'
- use personal contacts to 'persuade food-handling unions to support food irradiation as a residue-free process and invite union representatives to co-operate with the lead organizations'.
- engage 'a public relations company and/or communication consultancy to motivate food-handling unions, consumer organizations and print media to recognize the benefits and accept the application of food irradiation' and 'retain a reputable public relations firm schooled in "issues" management to achieve a positive response from [non-governmental agencies] and handle any negative issues that may be raised'.
- 'promote the general use of the logo [radura label] designed by the Netherlands as a symbol for irradiated commodities'.

Primary funding for this campaign,

> 'should be made available by the International Consultative Group to enable contact with the lead organizations in individual countries'.

In other words, taxpayers' money is to be used to finance the international campaign.

In addition, each of the UN agencies was asked to produce formal statements to be used in the promotion campaign.

- The World Health Organization — to declare food irradiation a significant method of reducing food-borne diseases
- The Food and Agriculture Organization — to endorse the importance of irradiation in post-harvest pest control and loss reduction
- The UN Committee on Trade Aid and Development, and the General Agreement on Tariffs and Trade — formally to declare that food irradiation can exercise a positive impact on international trade.[5]

There is clearly some very heavy international lobbying going on to gain acceptance and approval for food irradiation. Each acceptance or endorsement is used to help convince other organizations to endorse. Piece by piece the illusion of acceptance is fostered until it becomes a reality.

The strategy in action

Much of the marketing strategy has already been implemented. The World Health Organization has published a briefing on food poisoning and irradiation.[6] It has also published a book which was launched at an international conference on the acceptance of food irradiation held in Geneva, December 1988.[7] A video has been produced and a TV series is planned which will be offered to national networks around the world.

The International Atomic Energy Agency and the Food and Agriculture Organization have held many seminars and symposia on the contribution of irradiation to food safety and reducing food losses in different regions of the globe. They also have on-going 'research' projects, such as that for Asia and the Pacific. A meeting in 1988 was attended by representatives of the atomic energy/nuclear institutes of 13 developing countries in the Asia/Pacific region.[8] This discussed plans for 'demonstrating consumer acceptance' by carefully controlled marketing trials in several countries.[9] Concerns of consumer organizations are not allowed to interfere with this 'research'. At the meeting we were told, 'The only consumers who have a right to comment on irradiated food are those who have bought it!' Taken to its logical conclusion we suppose this same attitude would mean that the only people who have a right to comment on the dangers of nuclear war are those who have been in one. We wonder if the hazard of thalidomide would have been recognized if the only doctors permitted to comment on the deformities it produced were those whose own children were affected.

Any suggestion that outstanding issues should be resolved before the technology is promoted, particularly in developing countries, is met with hostility.[10] Rather than address the outstanding issues, the official reports are used to misrepresent the legitimate concerns of the international consumer movement.

The IAEA has a network of contacts in atomic energy institutes around the world. In developing countries, these institutes have been used as the lead organizations for promoting food irradiation. Because of the sensitive nature of work with atomic energy, these nuclear institutes are often exempt from close public or parliamentary scrutiny. In 1988, we intercepted attempts by the Philippines Institute for Atomic Energy Research to bypass the President, Mrs Cory Aquino, and establish a national committee for promotion of irradiation.[11] In Pakistan there have been attempts to establish a three-million person market for irradiated foods by supplying the Pakistan army. When it was pointed out that the Pakistan irradiator had been provided under a research grant from the IAEA which, as a UN agency, did not permit use of its funds for military purposes, it was announced that 'soldiers are consumers too'.[12] In China, there is no pretence. There is no need for promotion. Consumers will not be told that food has been irradiated. Irradiated food, we were told, will be labelled 'when there is consumer acceptance'.[13] In 1975 a petition of complaint was signed by over 100 of the 150 staff of the Office of Atomic

Energy for Peace (OAEP) in Thailand where the current head of the IAEA division promoting food irradiation then worked. The petition objected to the unlabelled sale of experimentally irradiated onions. The OAEP director resigned.[13a]

The public in many developed countries is more sensitive to the activities of the nuclear industry. In Europe and North America the ICGFI has used the existing food irradiation companies and a few large food processing companies as the focus for creating the 'lead organization' to spearhead promotion. In the USA a 'Coalition for Food Irradiation' had some 33 leading companies involved by 1986.[14]

In the UK, the initial promotion came from a group co-ordinated by the Food Industries' Research Association, based at Leatherhead in Surrey. In addition to the Research Association, the group consisted of representatives of:

- Isotron — the company with a virtual monopoly on gamma irradiation facilities capable of handling food in the UK
- Radiation Dynamics — the leader in electron beam and X-ray sterilization techniques
- Unilever plc — the UK/Dutch multinational food manufacturing company.

This group was initially successful in persuading the media to carry stories suggesting that both consumers and the food industry wanted irradiation.[15] About 60 food companies provided finance for research aimed at rebutting some of the safety and wholesomeness concerns. The Leatherhead group provided speakers in favour of irradiation for many UK meetings and media interviews. The pro-irradiation campaign was successful in persuading the Food and Drink Federation, representing the large UK food manufacturers, to finance a booklet promoting irradiation in 1986. The task of promotion has now been taken over by the UK government using taxpayers'[16] rather than industry money.

World-wide, the marketing strategy emphasizes the importance of safety assurances given by influential bodies such as the World Health Organization, national advisory committees and medical and scientific bodies. Scientific societies are to be persuaded to organize 'scientific seminars and press briefings'. The Royal Society in the UK organized a wholly one-sided 'briefing' in 1987. Unfortunately for the organizers, the meeting was attended by a representative of the poultry industry who pointed out that it was not possible to market chicken on the basis of telling consumers you have killed nine-tenths of the bacteria that would give them food poisoning. Journalists present commented on the relation between the meeting and the ICGFI marketing strategy.[17]

In many countries, particularly in the Third World, World Health Organization endorsement of the promotion campaign carries considerable weight with the local scientific and medical bodies. Legislators in these countries rely on such international and local scientific opinion. Few people have the time to look behind the assurances and establish the facts. Even in developed countries these assurances are often taken at face value. The American Medical Association states,

> Many years of international experience have demonstrated that food irradiation at levels up to one megarad (1 kGy) is a safe process. Food irradiation produces no significant reduction in the nutritional quality of food. Moreover, it has a number of important potential applications. Food irradiation holds the promise of being a viable alternative to the pesticide ethylene dibromide in the post-harvest disinfestion of fruits and vegetables. It may also be used to control Salmonella in red meats, poultry and fish. In addition, food irradiation could extend the storage life of numerous perishable foods.[18]

The statement could have been written by the promoters of irradiation rather than a national medical body. In fact the AMA admits that,

> The American Medical Association . . . has not conducted any studies or issued any reports on the safety of food irradiation. At the present time, neither the AMA's Council on Scientific Affairs nor the group on Science and Technology has any plans to conduct such a study.[19]

In the UK, the British Medical Association's Board of Science severely criticized the government's Advisory Committee Report in 1987.[20] In 1989 it publicly criticized the suggestion that irradiation would have any significant role to play in tackling the problem of food poisoning — arguing that it could make the problem worse.

Some aspects of the marketing strategy have clearly failed. In addition to the British Medical Association, a wide range of influential national and international organizations remain opposed to irradiation — at least until the controversial issues are resolved. These include several national governments, many local, state and provincial governments, parliamentary committees in several countries, the European Parliament, almost all consumer organizations world-wide, all environmental groups, a large number of women's organizations, most trade unions with members in the food industry, a number of public health bodies, food law agencies, and some significant sections of the food industry. A list of some of the major UK organizations critical of the UK decision to remove the ban is given in Table 11. A similar list of responsible and influential critics could be compiled for many other countries.

Table 11: Major UK organizations critical of the government's decision to remove the ban on irradiation of food (1989)

The British Medical Association
The National Consumer Council
The Consumers Association
The Consumers in the European Community Group
The National Federation of Women's Institutes
The National Council of Women
The National Farmers' Union*
The Retail Consortium
The Institute of Environmental Health Officers
The Transport and General Workers Union
The General, Municipal and Boilermakers Union
The Bakers Food and Allied Workers Union
The Union of Shop Distributive and Allied Workers
The National Union of Public Employees
The British Frozen Food Federation
The British Mushroom Growers Association
Friends of the Earth
Greenpeace
The Soil Association
The Vegetarian Society
The Food and Drink Federation *

**Not opposed but seeking detection methods first (1987)*

However, we cannot claim that the world has yet seen through the public relations strategy and decided to reject irradiation. How do the various forces line up?

Governments

As of 1989, only 35 countries permitted irradiation and fewer than 20 had irradiation facilities. Of the approximately 120 countries remaining, six clearly opposed widespread use of irradiation. These were Australia, New Zealand, Sweden, West Germany, Denmark (which permits only spices) and Japan (which permits only potatoes). Denmark and West Germany will be pressured to permit it by 1992 under an EC trade harmonization directive. Within the EEC, Ireland, Greece and Portugal did not permit it but are unlikely to strongly oppose it. The remaining six EEC countries permit it. In Scandinavia, Norway and Finland allow its use on spices only.

The promoters of irradiation suffered a major setback in 1988. After a week-long conference on the acceptance, control of and trade in irradiated food, Australia, New Zealand, Sweden, West Germany, and Denmark publicly dissociated themselves from the conference statement before the chairman hastily closed the conference.[21] Subsequently, Iran, Libya, Saudi Arabia, East Germany and Brazil also raised objections or reservations about the promotional statement.

In the USA three states, Maine, New York and New Jersey, have passed legislation banning the sale of irradiated foods. Similar legislation was introduced but not passed in a further eight states. Many local councils have also used their limited powers to oppose it. In the UK many local authorities oppose irradiation and several Environmental Health and Trading Standards Officers have been actively involved in our working group since 1985. Parliamentary and other consultative Committees in Europe, Canada, New Zealand and Australia have handed down reports highly critical of irradiation.

Parliamentary committees involve elected representatives who can call on both scientific expertise and other sources of information. The result is that they tend to approach the issue from a more common-sense perspective than the narrow one adopted by the scientific advisory committees that the promoters of irradiation rely on. In 1989, the Australian (all-party) parliamentary committee unanimously called for the World Health Organization to re-open the safety investigation and produce a proper report.[22] In 1987, the Canadian (all-party) committee also concluded there was not sufficient proof of safety.[23] The Canadian government ignored this report. A broadly constituted advisory committee for the New Zealand Ministry of the Environment concluded that irradiation was not needed and should remain banned.[24] A panel of lay people, set up in 1989 to advise the Danish Parliament, concluded that even the permit for spices should be revoked on the grounds of unresolved issues of safety, control and the damage that irradiation would do to Danish food culture — based as it is on an image of hygiene and quality.[25]

Consumers

There has been an almost complete failure to win over consumer organizations. In 1987, the International Organization of Consumers Unions (IOCU) unanimously adopted a position calling for a world-wide moratorium on any further use or development of irradiation until there is satisfactory resolution of a number of outstanding issues. (The full text of the IOCU position is given in Appendix C2.)

The IOCU is the UN accredited body representing the world's consumers. It would seem to be the obvious and legitimate body to address an international conference on food irradiation. However, in 1988 the UN organizers of the international promotion conference chose a private individual to speak on consumer acceptance. She tried to suggest that the organized consumer movement was unrepresentative of real consumers and that it had been 'hi-jacked by a vociferous anti-nuclear movement'.[25a] The paper was liberally sprinkled with insults and singularly short on facts. It is worth noting that the IAEA report of the conference omits parts of the speech which was actually given and then misrepresents the accreditation of the speaker in a footnote to the critical comments of the IOCU delegation. A small point perhaps but indicative of the fact that the promoters have no qualms about bending the truth to suit the needs of the promotion campaign.[25b]

The WHO turned to Mrs M. Young, a representative of the Canadian Consumers Association to help give a façade of legitimacy to its promotion booklet on food irradiation in 1988.[25c] Mrs Young is one of the few consumer representatives in the world who has supported food irradiation. The booklet Mrs Young helped to write presents a series of questions and answers in the section on 'Consumer Acceptance' that appear, to us, to divert attention from the real concerns of consumers rather than to answer them. We ask why so unrepresentative a person is used in this way when the world-wide consumer movement has clearly voiced its concerns.

The organizers of the 1988 international promotion conference initially tried to prevent consumer representatives from formally presenting their case. Under pressure, they scheduled a special session to discuss outstanding safety issues and abuses. The official response[26] inadequately acknowledges the priority questions raised by consumer representatives[27] and fails to address the interrelated nature of toxicological and nutritional effects (see Chapter 3). The organizers were unable to persuade the conference to agree to an official statement that there were 'no unresolved safety issues'.

It was a disturbing experience to find that it was the consumer movement which had to press the technical point on the need for hygiene standards for food prior to irradiation. The organizers, including the World Health Organization representatives, and most governments did not want mention of such standards in the conference statement. They argued that a statement that food should be handled according to 'good manufacturing practice' was sufficient — no detail was needed on how this should be interpreted. Indeed some delegates argued that good manufacturing practice meant the use of irradiation to reduce unacceptable levels of bacteria! Eventually, this issue was deferred to a special meeting on standards for irradiated food held in 1989.[28]

The final insult in Geneva came during the concluding press conference. The organizers, in some desperation it would seem, tried to suggest that consumer opposition to irradiation was based on confusion with radioactive contamination from the fallout after the Chernobyl accident! This is far from the truth. Consumer resistance has grown as the propaganda campaign to promote food irradiation has failed to address consumers' real concerns. Confusion with radioactive fallout is simply not an issue.

World-wide, concerns are being expressed by consumer organizations. These are reasoned concerns about the safety, quality, abuse and control of irradiation and some searching questions about whether it is needed to the extent that its promoters would have us believe. These concerns of consumers are based to a very large extent on the arguments and issues presented in this book. They are shared by a wide range of other respected and influential bodies. They demand honest responses from governments and the agencies whose mandate is given by the United Nations on behalf of the people of the world. Until these issues are resolved there will be no consumer acceptance of irradiation — there will be active consumer opposition to it.

Trade unions

In the UK, all five of the major unions with members in the food industry have actively supported our work over the past five years (see Table 11). These unions have also been actively campaigning within the European and international trade union structures. As a result, the European Committee for Food of the International Union of Foodworkers (ECF/IUF) voted for a Europe-wide ban on irradiation in 1987. The ECF/IUF report states,

> Techno-physiological processes should only be used when, first, they present no health hazards and secondly when they are technically indispensable. . . The ECF has considerable reservations in this field as to whether consumers really can be protected against insidious deterioration in quality and fraudulent practices.[29]

The Asia/Pacific region of the IUF representing the food workers in many SE Asian countries also voted to oppose irradiation in 1989. In the USA the Food and Allied Services Trades of the American Federation of Labour and Congress of Industrial Organizations (FAST-AFL/CIO) has also taken a critical stance. The Health and Safety Director of FAST, Debbie Berkowitz, has stated,

> It is unclear whether any benefits in this area outweigh the risks to workers and consumers. We are deeply concerned about the worker safety aspects and feel the whole issue

> needs further study before any employment of irradiation technology. In addition, our members being consumers and residents, we also have deep concerns about consumer risks from consumption of irradiated foods and transport of radioactive materials around the country.[30]

The Food Industry Union Federation of Australia has taken a leading role in the national coalition opposing irradiation in Australia.

Unions have been alerted to the marketing strategy of co-opting a few trade union leaders into promoting food irradiation. Almost universally the trade union movement has sided with the consumers, linking worker and consumer safety issues and arguing that their members are consumers too.

Public health and food law agencies

Britain is almost unique in Europe in having food law enforcement responsibilities at local government level. Local authority Environmental Health Officers, Trading Standards Officers and Public Analysts are on the front line of inspection and enforcement of food safety. Even at major international sea ports like Southampton, these functions are still performed by local government, paid for with local government money. This has both advantages and disadvantages. It means that the local authorities, with limited resources, have had to establish their own networks for sharing information about suspected abuse so that rejected food does not re-appear at another port after a brief trip to the Netherlands. It has, however, meant that local authorities have been able to review the problems of food irradiation free from the political control of central government. The result is that many of these local government food law services have become highly critical of the government's plans for food irradiation. Some have been involved in the research programme undertaken at the London Food Commission since 1985. In 1989 the professional body, the Institute of Environmental Health Officers, publicly recommended that the government should not remove the ban.[31]

In many other countries, food law officers are civil servants and less free to take a position critical of government policy.

Despite this, some professional bodies have taken a stand. In Australia, the Meat Inspectors' Association actively lobbied for the Australian government to retain the ban on irradiation.

The food industry

Attempts to convince the food industry of the need for irradiation have been only partially successful, despite the involvement of some large corporations. The Swiss multinational Nestlé has frequently been represented at meetings of the IAEA/FAO International Consultative Group. The Dutch/UK Unilever company helped to establish the promotional group based at the Food Industries' Research Association in the UK. However, there has been little support for the campaign to raise industry money to promote food irradiation.

There was an attempt in 1985 to use the media to create the illusion that the food industry and consumers were waiting for food irradiation with bated breath. In fact a survey we undertook found only one leading company in favour of irradiation and most organizations believing that there would need to be considerable increase in regulation and control if it were to be permitted.[32] The UK manufacturers' Food and Drink Federation (FDF) published a promotional booklet in 1986.[33] In 1987, the chairman, faced with a national opinion poll showing that 93 per cent of the general public did not want the ban on irradiation removed, stated,

> . . . I don't know what our position was before, but as of today we are not seeking to have the ban removed until there are methods for detection.[34]

The UK Frozen Food Federation has taken a clear stand against the potential for abuse of irradiation.[35] Various growers' associations[36] have expressed reservations, as have the National Farmers Union[37] and, particularly, the large supermarket chains. The Retail Consortium joined the Consumers in the European Community Group and the Institute of Environmental Health Officers in asking the government not to remove the ban in 1989.[38] Three of the UK's largest retailers, Tesco, Marks & Spencer and the Co-operative Society, have all declared they have no wish to use irradiation. Marks & Spencer stated that it would not fit within their policy of providing 'quality fresh food on a fast turnover'. The only food industry group clearly in favour of irradiation is the British Spice Trade Federation.[39]

In the USA, membership of the pro-irradiation Coalition for Food Irradiation has fallen from its peak in 1986. A number of major companies have withdrawn their membership. Several more have declared that their involvement was 'for information only' and that they had no plans to use irradiation.[40] A major campaign is underway in the USA to establish the policies of food companies on irradiation.[41] In Australia at least one of the leading retail chains suggested they had no use for irradiation. [42]

The UK National Farmers' Union has reservations about irradiation. There is little indication that the farming/agriculture sector of the food industry will need irradiation, certainly not in Europe where there is massive overproduction at the present time. In any case, most of the food grown now passes through the hands of large food processing companies and, increasingly, large supermarket chains on its way to the consumer. Clearly the stance of these sectors of the food industry is critical to the future of food irradiation.

In most developed countries, there has been a significant change in the structure of the food processing industry. Looking at the variety of foods on supermarket shelves, consumers can be forgiven for thinking that there must be thousands of companies involved. The reality is that the market in many food items is dominated by just two or three companies. In confectionery three companies, Rowntree, Cadbury and Mars, control 60 per cent of the market. The savoury snacks market is dominated by Pepsico (Smiths and Walkers), United Biscuits (KP) and Dalgety (Golden Wonder) who share 70 per cent of the market. Unilever (Birds Eye), United Biscuits (Ross) and Nestlé (Findus) share 64 per cent of the frozen ready-meals sales. Just two companies, Heinz and Campbell's, share 71 per cent of the canned soup market. Table 12 lists the major European food manufacturers.

Advertising is a key component of the industry's strategy to win sales and 'brand loyalty' from consumers. It also excludes potential competitors entering the field. To give some idea of the scale of priorities, the UK government has spent around £4 million on the health campaign 'Look After Your Heart', which places some emphasis on the role of diet in promoting or combating heart disease. In 1988, the food industry spent £570

Table 12: Major food manufacturers

Company	Activity	Nationality	Market capitalization approx (£bn)
Unilever	Foods, detergents, chemicals	UK/Netherlands	8.5
Nestlé	Food manufacturing	Swiss	6.5
BSN	Food and glass products	France	2.6
Cadbury Schweppes	Manufacture of confectionery and soft drinks	UK	2.2
Associated British Foods	Food manufacturing & retail	UK	1.3
Ranks Hovis McDougall	Manufacture, processing & retailing of food products	UK	1.3
United Biscuits	Manufacture & sales of food products	UK	1.2
Hillsdown	Food manufacture, furniture & property	UK	1.2
Suchard	Coffee and chocolate manufacturers	Swiss	1.1
Source Perrier	Mineral water	France	1.0
S & W Berisford	Food, financial services property & commodities	UK	0.8
Dalgety	Agribusiness, food manufacture & commodities	UK	0.7
Unigate	Manufacture & distribution of food, transport etc.	UK	0.6
Northern Foods	Manufacture of foodstuffs	UK	0.6
Tate & Lyle	Agribusiness, sugar refining, malting, distribution, insurance	UK	0.6
Booker	Agribusiness, health products, food distribution	UK	0.6
Beghin-Say	Sugar and paper products	France	0.6

Source: Financial Times/CL-Alexanders Laing and Cruickshank, November 1988

million mainly on TV advertising (see Table 13). Nearly a sixth of all advertising is spent on promoting food. Clearly, the money the industry has to spend on influencing consumer taste and, perhaps, consumer acceptance of irradiation is considerable.

Table 13: Amount spent on food advertising in 1988

	£ million
Chocolate	81.7
Cereals	61.2
Coffee	55.1
Sauces, pickles, salad cream	29.2
Margarine	25.5
Tea	25.1
Frozen meals	23.6
Potato crisps and snacks	20.9
Sugar confectionery	19.6
Biscuits	19.0
Fresh and frozen poultry and meat	16.5
Meat and vegetable extracts	15.9
Milk and milk products	14.7
Cheese	14.3
Butter	9.7

(Source: MEAL/Euromonitor)

Nevertheless, of the three main sectors (agriculture, manufacturing and retailing) the retailing sector is the most powerful. The growth of the large supermarkets has resulted in six major chains controlling almost 70 per cent of the market in food sales to the consumer (see Table 14).

Table 14: UK grocers' market shares 1987/88

Company	Market share (per cent)
Tesco	14
J Sainsbury	13.9
Dee (Gateway/Fine Fare)	11.5
Argyll (Presto/Safeway)	10.7
Asda	7.6
Co-op	12.1
Total	69.8

(Source: Verdict Research/*Financial Times*/*Food* Magazine, Spring 1989)

Retailers claim they mediate between the consumers and the producers and processors. However, retail power lies in their ability to specify exactly what will go into the products. Even the largest manufacturer has to comply if it wants its product on the supermarket shelves. Yet it is the supermarkets which are most susceptible to consumer pressure. Supermarkets are already responding to consumer demand on a number of issues. The recent trend towards 'additive-free', 'natural' and 'ozone friendly' products suggests that some supermarkets have decided consumer feeling on some issues is so strong that they need to respond. Thus it is the supermarket chains which probably hold the key to acceptance or rejection of irradiation, provided that consumer concerns can be clearly shown to be opposed to it.

All the costs of processing, packaging, transporting and advertising food are eventually passed on to the consumer. So will the additional costs of irradiation. The World Health Organization quotes a range of costs between US $0.02 and $0.40 per kilogram — roughly half a UK penny to ten pence per pound weight.[43] These costs may be optimistic. The Italian Chapter of the International Organization of Consumers Unions has calculated that, spread over a five-year period, the total costs of irradiation could add £50-60 to the cost of a ton of potatoes.[44] The key element is not so much the running costs as the very high initial cost for construction of an irradiation plant. This cost has been variously estimated at between £1 million and £5 million. American designs were priced at $1.1 million to $2 million in 1982/4.[45] The US Department of Agriculture estimates costs of between $1 million and $11.2 million.[46] However, the USDA admits that the total costs involved in operating a plant are unclear. These will depend on such factors as:

- *the seasonal availability of food.* Ideally, irradiation facilities need a constant supply of a single raw material. Changing conveyor speeds, different handling requirements and 'down time' when the plant is not operating (but the radioactive source is still decaying) all add to the costs.
- *the cost of the radioactive source* — the cost of cobalt 60 or caesium 137 or energy costs if it is a machine source. These vary considerably and, as we will show, involve some hidden subsidies.
- *location of the facility* in relation to the food supply and

potential markets. Transport costs to and from the facility can significantly alter the economics of irradiation.
- *the size of the facility.* In general there are economies of scale but only if the availability of a constant supply of locally produced and marketable food can be guaranteed.
- *consumer acceptability,* which can have significant impact on the market for the irradiated food and hence the economies of scale.

Long-term hidden costs are also involved. These include the effects of structural changes in the industry such as those leading to greater concentration of power in the hands of even fewer companies, and the effects on employment within the food industries. These costs are eventually borne by the whole community. A full study of the economics of food irradiation would consider these factors. Just as hidden subsidies can make the process appear profitable, hidden costs could indicate that it is undesirable. A study taking all factors into account is needed to help assess the claims made about the economic benefits of food irradiation.

The food irradiation industry

One group clearly stands to benefit from a global campaign to promote food irradiation. This is the food irradiation industry itself. There are also companies irradiating medical supplies, wool, carpets, plastics and animal feedstuffs which might see a bonanza if irradiation were permitted and the food industry and consumers could be persuaded to accept the process. With the high costs and inevitable lead times before competitors could enter the field, these firms stand to profit enormously if there is a rapid take-up of food irradiation. In Europe, the USA, Australia and elsewhere, these companies have been actively promoting irradiation. Many of their public statements on the benefits of irradiation appear to be made as much with an eye on the stock market value of the companies as to the effect on public opinion. People associated with these companies have also been active in influencing government decisions on irradiation. Despite the obvious conflict of interest, some of the key players have been invited to take part in preparing government reports and in planning promotion activities organized through the IAEA/FAO International Consultative Group.

In the UK, Isotron plc was clearly the major company which would benefit from a change in the law on irradiation. Many of the early public statements on irradiation were made by Frank Ley, Marketing Director of Isotron. Ley was actively involved in setting up the promotional working group on food irradiation in 1985. Statements on the benefits of food irradiation were linked throughout 1985 to hints that the UK government Advisory Committee was about to approve the process. Isotron had spare production capacity of around 46 per cent at its existing plants.[47] Frank Ley had previously worked in the food research department at Unilever and as principal scientific officer at the United Kingdom Atomic Energy Authority, leading a team investigating the irradiation of food. In 1970 he left to set up the companies which eventually merged to create Isotron.[48] In September 1983 Ley, now the Marketing Director and a leading shareholder in Isotron, was appointed 'economic advisor' to the UK Advisory Committee.

Several British MPs tabled a motion in the House of Commons in 1986 indicating their concern over possible conflict of interest (see Appendix B1). Specifically, they noted that:

- predictions of the main recommendation of the Advisory Committee had been widely leaked, not least by Frank Ley
- the company had raised capital to build a new irradiation plant through a flotation on the stock exchange while the committee was sitting (despite overcapacity at existing plants)
- there had been a rise in the capital value of Isotron when stories in the financial press linked the future of the company to the impending recommendations of the Advisory Committee.

The motion called for an investigation of share dealings in the company.[49]

Also around this time it was revealed that Sir Arnold Burgen, the chair of the Advisory Committee, was a part-time director of Amersham International, Britain's leading isotope manufacturer.

Clearly, it does not help public acceptance of the impartiality of the Advisory Committee report to have suggestions of conflict of interest. This would be of little consequence if the scientific evidence put forward by the report was impeccable; however, as we have shown, it was not.

A similar pattern emerges from analysis of the financial press in the USA. Between 1983 and 1986 the future profits of two of the leading irradiation companies, Radiation Technology and Isomedix, were linked with impending approval of irradiation by the Food and Drug Administration.[50]

- Various reports quoting Martin Welt, President of Radiation Technology throughout 1983, hinted that the FDA was about to give clearances for foods. In fact, Radiation Technology was petitioning for permission to irradiate these foods.
- In July 1983 Isomedix raised around $16 million capital through a share flotation to finance the cost of constructing an additional three irradiation facilities.
- In April 1984 the Securities and Exchange Commission of the New York Stock Exchange uncovered insider trading involving the shares of both Radiation Technology and Isomedix linked to favourable reports on both companies in the Wall Street Journal.[51]

Martin Welt, though not implicated in illegal share dealings, was required to stand down as president of Radiation Technology Inc. in 1986 by the Nuclear Regulatory Commission — to be promptly hired by the US Department of Energy programme promoting food irradiation. As noted in Chapter 5, Welt was convincted and sentenced in 1989 over deliberate violations at one of his company's plants. He now runs the Alpha Omega irradiation company promoting irradiation in SE Asia. George Giddings, formerly of Isomedix, now promotes irradiation in South and Latin America.

In 1986 Radiation Dynamics, the British X-ray technology company also involved in the British working group for food irradiation, announced plans to raise capital by share flotation. Clearly the irradiators are on to a good thing.

Frank Ley of Isotron was invited to the International Consultative Group meeting at Cadarache, France, in 1988 to help preparations for the further international promotion of food irradiation. He returned home for personal reasons after discovering that we were at the meeting.[52] Jan Leemhorst, of the Gammaster Company which has been implicated in many of the 'Dutching' of food scandals, was invited to the July 1988 ICGFI meting in Vienna and was also present at the Geneva conference promoting international acceptance of food irradiation in December 1988. George Giddings was included in the official US delegation to this conference. It seems that those with clear vested interests have easy access to the international machinery established for promoting food irradiation.

The plain fact is that pressure to permit irradiation came originally not so much from the food industry as from the food irradiation industry. This can be seem from the press reports in the UK in 1985/6. The US Food and Drug Administration states that, in the USA:

> The food industry has an interest in the process but they are very cautious. Pressure to permit irradiation is coming from the irradiation industry.[53]

So far, most of this pressure has operated behind the scenes or through attempts to manipulate public opinion using the media. Now, however, the promotion has been taken over by other much more powerful interests.

In the past there has been little financing of promotion of irradiation by either food or food irradiation industries. The costs of the next phase of the campaign are being borne largely by governments and international bodies — which ultimately means by the general public. Which agencies are able to command such support within the treasuries of nation states?

A nuclear industry connection?

To understand the international forces pushing food irradiation, it is necessary to go back to the origins of the technology in the 'atoms for peace' programme of the 1950s. The International Atomic Energy Agency is the UN agency charged with the promotion and control of civilian nuclear technology. Its objectives are to find peaceful uses for atomic power and prevent diversion of nuclear materials into production of nuclear weapons.

Finding peaceful uses of nuclear energy is becoming increasingly difficult. Public disillusionment with both the safety and the economics of nuclear-generated electricity, particularly following the Chernobyl nuclear accident in the USSR, has not helped. A significant part of the work of the Joint Division of the IAEA and FAO was in the area of plant breeding. Some 40 years of work, attempting to use irradiation to produce mutations in plant species, has been made redundant with the advent of genetic engineering techniques. The new bio-technologies can reproduce useful genetic traits far more precisely than the hit-and-miss approaches of irradiation. Food irradiation is not merely another use of atomic energy, it is perhaps the last hope of a dying technology, seeking to find some crumb of public acceptance that will justify the enormous sums of money invested in it over the past 40 years.

Public acceptance of irradiation would provide several important benefits for the promoters of the nuclear dream.

Radioactive waste — the military connection

It would provide a commercial outlet for some problematic nuclear wastes from the atomic reactor programme. Finding a 'use' for caesium 137 would relieve some of the problems of radioactive waste management.

Use of caesium would also legitimize re-processing in order to separate this isotope from other radioactive materials in spent nuclear reactor fuel. Re-processing is under threat because of the possibilities it creates for diversion of another radioactive material, plutonium, into the production of nuclear weapons. Plutonium is currently separated both for weapons and for possible future use in fast-breeder reactors in the USA, USSR, UK, and France and, on a smaller scale, by the other nuclear weapons states and those seeking to acquire such weapons. Many countries are, however, abandoning their fast-breeder reactor programmes. If this continues it will be harder to hide the military re-processing behind the supposedly civilian programme.

The USA has banned the re-processing of fuel from commercial nuclear power reactors. This was ordered by President Carter in 1975 in an unsuccessful attempt to set an example that would prevent that proliferation of nuclear weapons. The USA does have a military re-processing programme run by the Department of Energy (DOE), with facilities at Hanford in Washington State and at Savannah River in North California. The commercial use of caesium obtained from this military programme is being promoted by the DOE under what is called the Advanced Radiation Technology Program (formerly called the By-Products Utilization Program). The DOE allocated $5 million a year in 1986 and 1987 for construction of demonstration food irradiation plants, all using caesium 137. The caesium costs $2 per curie to extract and will be leased at 10 cents a curie — i.e. one-twentieth of the cost. The supply of caesium from military re-processing is limited. Military reactors for plutonium production are also being shut down. Opponents of food irradiation in the USA are concerned that the DOE programme may be used as a way of persuading the US Congress to permit re-processing of the large stocks of civilian fuel. It could thus be a 'back door' way of obtaining further supplies of plutonium for the US arms programme.[54]

Atomic reactor markets

Food irradiation based on cobalt 60 would guarantee a future market for reactors producing industrial and medical isotopes. The Canadian government has privatized the state enterprise, Atomic Energy of Canada Ltd, which is the leading producer of cobalt 60. Now operating as Nordion, this company is aggressively marketing cobalt for food irradiation, particularly in developing countries such as Thailand, Jamaica and Malaysia. It is being assisted by the government with money from the Canadian International Development Agency.

The nuclear dream

Above all, any widespread acceptance of irradiation would demonstrate that nuclear technology can provide some public benfits after all. We have nothing but respect for those for whom the nuclear dream grew out of a commitment to turn the technology of the atomic bomb to peaceful purposes. We do however suggest that this dream is, in part, a mirage — slipping further away with each passing decade as more countries acquire nuclear weapons on the back of so-called 'peaceful' technology. We also observe that nuclear technology has failed to live up to its promises and has been a very poor investment. It has delivered very little in return for the billions spent on it. Its day is past.

It is possible to irradiate using machine sources. This would avoid many of the occupational and environmental risks associated with handling radioactive materials. This would separate irradiation from the nuclear energy and weapons programmes. Indeed, there are many technical and economic advantages to doing so. Machine sources can use higher energies, thereby reducing the wide differences between maximum and minimum doses to the food. They can be switched off when not in use and so, unlike radioactive sources, do not need continuous operation and large guaranteed throughput. They are also cheaper to operate. A Canadian study compared the total costs for construction, operation and maintenance of irradiation plants using X-rays, (Canadian) cobalt 60 and (US DOE subsidized) caesium 137. X-ray technology turned out to be 25 per cent cheaper than caesium and 50 per cent cheaper than cobalt.[55] If the other problems we have identified with irradiation can be resolved then, perhaps, this is the way that food irradiation should be developed.

A solution to world hunger?

Is irradiation any part of the solution to the problem of food losses in the developing countries, or is this another argument just being used to promote irradiation? Who would really benefit?

The World Health Organization states:

> the need to reduce food losses is paramount . . . However, developing countries will not be prepared to make the sizeable investment required to set up irradiation plants until the developed countries that import food give their approval to marketing food treated in this way.[56]

Is irradiation about increasing the marketability of cash crops — food that the developed world can afford — or about reducing losses of staple foods that poorer countries need? The WHO would suggest both, but if so where does the priority lie? It was noticeable at the IAEA regional project meeting in Asia in 1988[57] that the real drive to experiment with irradiation of rice came from those countries with surpluses to export. The countries which imported rice were singularly unenthusiastic about including this staple food in the list of food items to be 'researched' during the next phase of promotion.

Storage losses for cereals and pulses are estimated to be at least 10 per cent and for non-grain staples around 10-50 per cent. The loss of dried fish is around 25 per cent. In Ghana, for example, between 20 per cent and 30 per cent of storable foods are lost because of contamination with the *Aspergillus* fungus.[58] However, the real extent of food losses needs further research and it is by no means clear that many of these losses could be reduced by irradiation. It may be technically possible, but in practice it is unlikely. Most food production in developing countries is decentralized and food is often consumed locally. What would be the additional infrastructure costs — the roads, railways, bridges, warehouses, and other facilities needed for developing countries to be able to use irradiation on a significant portion of their food supply? What would be the effect on a rural economy of a large portion of the food supply being channelled through big corporations? Much of the problem of hunger is related to poverty rather than shortage of food. What would be the social and economic costs of using irradiation to reduce food losses?

A report for the European Community in 1987[59] recommends that thorough socio-economic feasibility studies should precede the introduction of food irradiation in developing countries because of the possible changes that will result in local food supply and marketing practices which may actually disadvantage the local population. It suggests the following factors need to be considered.

- Irradiation technology is capital intensive. It also requires a better infrastructure.
- This infrastructure allows large quantities of food to bc collected more efficiently for irradiation, but it also means that more produce can be collected for export. Developed countries will therefore benefit from having access to a wider range of produce.
- On the other hand, staple foods, that might be irradiated are those that are produced more easily in western countries.
- The increased purchasing power of developing countries from greater exports of exotic produce is therefore likely to be recouped by the richer countries' greater competitive

power in other areas of food trade and especially their ability to off-load surplus food.
- The industrialized countries are therefore likely to benefit most from introduction of irradiation in developing countries.
- For many developing countries, the construction of the infrastructure would equal the efforts necessary for the development of the entire country.
- Within the developing countries, the impact of irradiation on rural populations has not been fully thought out. For example, only regionally concentrated crops are worth considering for irradiation. Again, these are usually export products.

All this is far from the dream of solving the problem of world hunger. Indeed, it is reminiscent of much of the criticism of the so-called 'green revolution' which increased crop yields in many developing countries, but at a price. The price paid included:

- greater dependence on the products of the multinational agribusiness corporations for high-yield seeds, and the pesticides and fertilizers without which these crops cannot be grown
- increased personal and national debt to the international banking institutions, bankruptcy among small producers, and increasing concentration of productive land in the hands of wealthy landowners.

Whatever the benefits of the green revolution — and there were some — there were many problems as well. The result is that many people still go to bed hungry. The dream of using irradiation to relieve world hunger turns into a nightmare when one thinks of the food mountains in Europe and North America being irradiated in order to be stored longer, and these nutritionally depleted foods being off-loaded into the Third World — with insult added to injury by calling it 'aid'.

The people in the developing countries say (when asked) that their need is for appropriate technologies [60] such as:

- cardboard boxes to reduce the damage caused to produce in handling and transporting it from field to market

- simple, cheap and efficient transport vehicles that can be maintained and repaired under local conditions
- basic roads and rail services
- storage facilities that keep food off the ground and under cover, keep insects and rodents out, and control temperature and humidity so that aflatoxin-producing fungus and other moulds do not develop.

Irradiation comes a long way down this list. If the promoters of irradiation are serious about the problem of food losses then let them propose a comprehensive programme of development putting these other facilities in place first. It can then be seen whether irradiation is needed.

In Chapter 5 we analysed the economics of using irradiation to deal with food poisoning. The cost of irradiation is so high that the benefits need to be very large. There are other, more appropriate means (such as better hygiene) for reducing the incidence of food poisoning. This means that there is considerably reduced benefit to be gained from using irradiation — it simply is not economically justified. A similar situation applies with food losses. When other, more appropriate measures are used to reduce the extent of food losses, the economic case for irradiation is much weaker. Perhaps that is why it is being promoted as one of the first rather than the last technologies to deal with this problem of world hunger.

A food irradiation campaign?

This then is how the global battle lines have been drawn. The International Atomic Energy Agency, with some support from the Food and Agriculture Organization and the World Health Organization, is actively promoting acceptance of irradiation by governments, the food industry, non-govermental organizations and consumers. The UN agencies are being assisted by some large food manufacturers, the existing food irradiation industry, nuclear/atomic energy institutes, companies selling radioactive materials, some governments and the military establishment of some countries.

Use of irradiation of food is opposed by a wide range of organizations. These include:

- the International Organization of Consumers Unions representing some 170 consumer bodies in 70 countries including all the leading consumer organizations in Europe, Japan, South and South-east Asia, Australasia and Latin America
- the food-workers section of the international trade union movement in Europe, North America, and the Asia/Pacific region
- the Environment Liaison Centre International and all the leading environment groups world-wide
- sections of the food industry, including major international traders in products such as seafood, large retailers and a number of national farmers organizations
- public health and consumer protection agencies including the UK Institute of Environmental Health Officers and the British Medical Association
- the governments of Australia, New Zealand, West Germany, Denmark and Sweden. A Canadian parliamentary report opposes irradiation in Canada and the Australian Parliamentary Committee has called for the World Health Organization to re-open the investigation of the safety of irradiated food and produce a properly-referenced scientific report. The European Parliament voted against a general clearance for irradiation in 1987 and voted for a general ban on irradiation (with an exception for spices) throughout Europe in 1989.

This opposition from responsible and respected organizations cannot be dismissed, as some in the pro-irradiation lobby have asserted, as simply an uninformed or emotional reaction.

Irradiation is also opposed by many ordinary consumers. In the final analysis it is the views of the consumers which will count. As the promotional conference in Geneva in 1988 conceded, 'no technology can be applied unless there is consumer acceptance'.

National surveys and opinion polls in the UK in 1986, 1987 and 1989 have shown that 70-85 per cent of people asked were decidedly opposed to buying irradiated food. Only 13-20 per cent indicated a willingness to buy it provided there were assurances on safety, and only 25 per cent would trust government assurances on safety. If it is permitted, 95 per cent want all irradiated food labelled and 93 per cent want a method for detecting whether a food has been irradiated before the ban on irradiation is removed.[61] Similar results have been obtained in the USA.[62 63 64] Furthermore, studies in various countries have found that resistance is greatest among the richer and more educated consumers and is stronger when more information is provided about irradiation — even when this is provided by promoters of food irradiation.[65 66 67]

The stage is set for a major struggle to determine whether irradiated food will be forced on to consumers before all the unresolved issues have been settled. Consumers need to be concerned about the role of the UN agencies and governments in the promotion of food irradiation, not least because of the extensive use of public funds being used for this purpose. The promotion campaign is diverting resources from the scientific research that is still needed. Failure to resolve the outstanding issues may prevent us gaining any benefits there might be from technology.

A few years ago when we began our study of food irradiation, we considered ourselves open-minded critics. We have now become opponents. We have done so reluctantly because all criticism and concern has been either ignored or brushed aside. Irradiation is to be 'accepted' by the world's people whether they like it or not. If commonsense, concern for public safety, and the right of the public to participate in the making of decisions such as these are to prevail, then irradiation must be actively opposed. Only then will the issues be addressed. Only then will it be possible to have whatever benefits there may be from irradiation — without the unwarranted risks.

7. What is to be done?

There does not appear to be any real need for irradiation — certainly none that would justify overriding the real concerns we have raised in this book. Nevertheless, its use is being promoted within the corridors of power, within the boardrooms of industry, and using the various avenues by which public opinion can be influenced — the media, advertising, and what the international marketing strategy calls 'controlled communications' — in short by propaganda.

Should we just accept this as inevitable? If the promotion campaign is being co-ordinated internationally, what chance does one country like Britain have of banning the process? What chance has Denmark or any Western European nation, Poland or any Eastern European nation, or any of the smaller, poorer, less powerful nations in Africa, Latin America or South-east Asia? Irradiated food, we are assured, will be labelled. Is that not enough? Should not people be left to choose whether or not to buy irradiated food?

We think not. First, irradiation of our food supply is far from inevitable. Second, this is not simply a consumer choice issue — it is a public health issue. Public health, at least in our society, is an area where the community intervenes to protect the individual from being manipulated by the advertising industry and the economic forces of the market-place. There is no advantage to the food industry in irradiating good quality food and adding to its cost. Irradiation will be used on lower quality food — to counterfeit quality.

A food quality campaign

Good quality food is what the rich can already afford. In the words of an old English song, 'It's the rich what gets the gravy and the poor what gets the blame!' We believe that blaming the victims of poor diet is not helpful. What is needed is a public food policy that seeks to improve the quality of the food everyone eats. In any case, in terms of the money spent as a proportion of their income, the poor do a much better job in providing for their dietary needs than do the better off.[1]

A new food policy for the 1990s would involve significant changes:

- *in agriculture* — using the opportunity created by massive food surpluses to make a transition to less intensive production, reducing the use of chemical fertilizers and pesticides that are so damaging to the quality of the soil, water, food and eventually our health.
- *in food manufacturing* — less use of low-quality over-processed raw materials made attractive by artificial additives, colourings and flavourings; greater use of good quality ingredients. Convenience and nutritional quality are not incompatible.
- *in retailing* — extending what many retailers are already doing; that is, responding to consumer-led demand for a wider variety of quality foods including fewer additives, less fat, salt and sugar, and fewer pesticide residues.

Comparing supermarket shelves today with those ten years ago shows that some of the changes we are seeking have already taken place. Good quality food can be processed for convenience and still be in line with the kind of dietary advice issued by national and international health organizations. The World Health Organization goal of reducing fatal disease by dietary change in Europe by the year 2000 is not simply a good idea. It represents a major saving in health-care costs which are paid for by the whole community, not just the individuals affected.

The revolution in dietary health needs to be extended to the disadvantaged, those who are old, in ill-health, disabled and poor. It needs to reach those of different cultures in other countries around the world. We need a new global agricultural revolution which learns from the lessons, mistakes and failures of the so-called 'green revolution' to tackle the real roots of world hunger.

Irradiation has no part to play in this revolution. It is, at best, irrelevant. At worst, it will be used as a high-technology fix for problems that have their roots in the way that the food industry currently operates. It will cover up for lower quality food and unhygienic food-processing practices. It is being promoted in developing countries where there is even less need for it, even more scope for abuse and where governments and public opinion can be more easily manipulated. It needs to be opposed in much the same way as the marketing of processed baby food to developing countries.

Thinking globally, acting locally

The choice is no further away than the nearest corner store, supermarket or street trader. The bottom line is that, as the World Health Organization admits,

> No treatment of food can be employed in the long term unless it has the acceptance of the consumer.[2]

Consumers have not accepted irradiation. What is important is that this message is conveyed clearly to each of the relevant institutions which influence public opinion and, ultimately, government decisions. The ways this can be done will vary from country to country but the pressure points and the message are the same. But, and this is the major lesson we have learned in the past five years, there has to be collaboration between people in different countries. The lessons learned elsewhere are relevant wherever you are. The actions you take can help people on the other side of the globe. The examples below indicate some of what can be achieved and some of what remains to be done.

Pressure on government

Governments enact food laws which either permit or ban irradiation. Governments also finance the international agencies such as the IAEA which are actively promoting the global acceptance of irradiation. One objective has to be to persuade governments to halt further development of the international propaganda campaign and to ban the use of irradiation until the various outstanding issues are resolved.

Governments in some countries, such as Australia, New Zealand, Sweden, Norway, Finland, Denmark, East and West Germany, Brazil, Iran, Libya, Norway and Saudi Arabia, have either banned irradiation, limited its use to a single food item, or called for further research into safety and methods for detection before it is permitted. Others, such as Japan, effectively ban imports of irradiated food. In the majority of the other countries in the world, the governments have yet to decide. It is permitted in only 35 countries and not actually used in many of these.

In the UK and most of Europe, however, irradiation is either already permitted or will soon be required under a European Commission Directive. This does not mean that writing to one's Member of Parliament or Member of the European Parliament is pointless. There is still a need for the public to inform both MPs and MEPs about the controversy over the safety of irradiated food, the problems of abuse and ineffectual controls, and the extent of public concern. Letters to MPs are never wasted. A single personal letter from a constituent draws attention to an issue. Ten letters suggest that maybe several hundred people in the constituency are concerned. Many letters signal that it is a public issue which may require the MP to consider his or her position and perhaps defend it in the constituency. Almost everything connected with the local MP is news in the local papers. An MP can draw attention to particular concerns by supporting both partisan and all-party initiatives. These are a way of indicating to government the strength of public outrage at the use of taxpayers' money and the tools of the advertising industry to promote irradiation.

There is also considerable scope for influencing elected representatives at local government level. Food law enforcement in the UK is in the hands of local government. Raising problems of abuse and detection of irradiation with representatives on Environmental Health/Port Health and Consumer Protection/Trading Standards committees has resulted in many local authorities opposing irradiation. Some have lent support to the campaign to keep it banned and have provided information on irradiation for local ratepayers. Some have decided that irradiated food will not be sold through local authority catering or used for school meals. Letters and visits to elected representatives are particularly effective if done by a group of people with specific suggestions for action they want the local council to take.

In the USA there has been vigorous lobbying at state and local level, particularly in areas where irradiation plants were about to be built. As a result:

- Not one of the six demonstration irradiators planned in 1986 had been started by the end of 1989.
- By 1989, three states — Maine, New York and New Jersey — had passed legislation prohibiting the sale of irradiated food within the state. Similar bills had been introduced in Vermont, New Hampshire, Massachusetts, Pennsylvania, Minnesota, Oregon, and Washington. California was to be added to the list in 1990.
- Over 100 federally elected representatives and senators supported legislation to prohibit irradiation of fruit, vegetables, pork and other foods and to require fresh scientific study of the effects of irradiation on human health and the environment.
- Test-marketing trials of irradiated papaya and potatoes had to be abandoned because of consumer lobbying and boycotts. The stores subsequently gave undertakings they would not sell irradiated foods.
- An international petition has been launched with a target of one million signatures — 350,000 have already been collected. The signatures are being used to persuade major US food companies, particularly the supermarkets, to guarantee they will not use or sell irradiated food. Many local stores carry a poster similar to that in Figure 4 stating, 'this store will not knowingly sell irradiated food'.
- Many of the food companies involved in the Coalition for Food Irradiation have now withdrawn their membership.

These initiatives have been backed by coalitions of community, public interest, consumer and environmental groups which have organized petitions, phone-ins, door-to-door canvassing, letter-writing campaigns, lobbying of legislators, boycotts of stores and demonstrations in support of their campaigns. All this in a country that legally permitted irradiation in 1963 and extended the permits in 1986. The result is that, as of 1990, no irradiated foods were openly on sale in the USA. Only ingredients such as spices and some wheat flour were being

Because our customers' wishes are more important than the shelf-life of our food

THIS STORE WILL NOT KNOWINGLY SELL IRRADIATED FOOD.

Exposing food to radiation may prolong food shelf-life. But issues of safety, wholesomeness, controls on and need for irradiated food have not been resolved. The management of this store will not knowingly expose its customers to irradiation of food. For more information, contact: London Food Commission, 88 Old St. London EC1V 9AR.

Figure 4: Poster displayed by stores with a commitment not to sell irradiated food

irradiated and sold, unlabelled, without consumers being able to express their choice.

The food industry

The food industry is vulnerable to consumer pressure, particularly the companies owning large supermarket chains.

In the UK, the National Farmers' Union and a number of growers' and traders' associations have already declared their opposition to irradiation at least until the key issues are resolved. So too has the Retail Consortium, representing the large supermarket chains. Many of the supermarket groups have taken clear positions. Marks & Spencer declared in 1986 that irradiation had no part to play in its policy of providing quality fresh food on a fast turnover. Tesco, Britain's largest retailer, stated a similar position in 1987. The Co-operative Society has taken a leading role in co-ordinating opposition to irradiation within the retailing sector in Europe. Waitrose will not sell irradiated food until satisfied about its quality and Gateway has said it will not rush into the sale of irradiated food.

On the other hand, some retailers appear to be luke-warm in their opposition to irradiation. A 1988 news article suggests that Sir John Sainsbury, head of the giant Sainsbury's supermarket chain, may personally favour irradiation (see Appendix C4). Customers of J. Sainsbury plc may wish to establish exactly what the company policy is.

The Frozen Food Federation took a cautious position on irradiation in 1986. The leading poultry producers have also been cautious, and the poultry federation has suggested its members have no plans to use irradiation. The British Spice Trade Federation appears to be in favour of irradiation and has indicated this to the government. Some multinationals such as the UK/Dutch giant Unilever or the Swiss Nestlé companies appear to be at the forefront of promotion of irradiation. The position of the other major food manufacturers is less clear. In the UK, these companies are represented by the Food and Drink Federation (FDF). The FDF and many of the member companies receive technical advice from the Food Industries' Research Association at Leatherhead, whose director, Alan Holmes, has taken a leading role in promoting irradiation. The FDF produced a one-sided booklet promoting food irradiation in 1986, but appeared to reverse its position in 1987 saying that detection methods were needed if irradiation was to be permitted. Since then, the FDF director has indicated personal support for irradiation. The organization appears to be happy to have the government do the work of promoting irradiation with taxpayers' rather than industry money.

Consumers have a right to know which of these large companies, many with brands that are household names, are intending to use irradiation. Late in 1989, a coalition of UK consumer and environmental groups launched what has been called a 'positive list' campaign. Similar list campaigns are being launched in other countries world-wide. The objective is to provide the public with a list of the companies and the branded products where clear assurances have been given that irradiation has not been used on any of the ingredients. As well as writing to companies to ask them to give such an undertaking, consumers are also encouraged to write and congratulate those on the list for their stand and to express their concern to others which are not on the list.

Many food products are traded internationally. Among these, seafoods and spices are clearly targeted for irradiation — and such use is regarded by many consumers as a clear indication of inadequate attention to hygiene on the way to the supermarket shelf. European consumer groups opposed to irradiation have joined with those in the USA, Canada, Latin America, Australia, New Zealand, Japan, India, and South-east Asia in an international seafood campaign. The campaign will alert the producers of shrimp and prawns, mainly in Asia, of the damage that is being done to the image of seafood products from their countries by 'Dutching' of these foods in Europe. Even greater damage will be done if the Asian countries permit irradiation of these food products.

The simple fact, that should be obvious to anyone who has read this far, is that good food doesn't need irradiation. The food industry world-wide needs to realize that if any food has had to be irradiated consumers will ask what was wrong with it. The national and international campaigns will provide positive incentives to food companies to provide good quality, clean food and clear assurances that they will not use irradiation.

Non-governmental organizations (NGOs)

Consumers', women's, environmental, community, health, trade union and other groups are important vehicles for raising public awareness about food irradiation. The promoters' strategy has been to co-opt support from these as a way of establishing the legitimacy of its claims for the benefits of irradiation. This has not worked. In the UK many of these non-governmental organizations have opposed removing the ban on irradiation. Similar coalitions of NGOs are campaigning against irradiation in other countries. This has come about largely as a result of the action of individual members of these groups.

Almost any organization could be used by the promoters of irradiation. One example is the apparently generous offer by the US irradiation company, Radiation Technology, to supply irradiated food for Operation Raleigh in 1986. Operation Raleigh is a charitable project involving teams of young volunteers from the USA and Europe working on community projects in developing countries. Our US colleagues obtained copies of correspondence between Radiation Technology, Operation Raleigh, the US Food and Drug Administration and the US Department of Agriculture through the US Freedom of Information Act. This Act provides public access to much information which, in Britain, would probably be covered by the Official Secrets Act.[3] The irradiated food offered included items that would not have been permitted to be irradiated in the USA (or in many European Countries either). The US authorities readily gave permission for the irradiated food to be exported for consumption by the young volunteers provided it was 'legal' in the countries where they were working. The Patron of Operation Raleigh is HRH The Prince of Wales. We were concerned at an apparent attempt to use the young volunteers and Prince Charles in a public relations exercise. We supplied details of the correspondence to Buckingham Palace. Operation Raleigh declined Radiation Technology's offer.[4] How many organizations have unwittingly been used in such irradiated food propaganda stunts? It is claimed that US astronauts have eaten irradiated food. Conferences on irradiation regularly provide irradiated meals. Much propaganda use is made of the UK permit for irradiation of the diet of medical patients — despite the fact that the only hospital which we have been told was using this approach discontinued its use several years ago. Attempts are being made to promote irradiation for aircraft meals. British Midland Airways published a promotion article in the company's flight magazine given to all passengers.

Organizations concerned about food irradiation have been linking up across national boundaries.

- A network of European consumer and environmental groups was established in 1986. Co-ordinated lobbying led to the European Parliament vote rejecting a general clearance for irradiation in 1987 and opposition to the general European Directive on food irradiation in 1989.
- Meetings have been held in the Asia-Pacific region involving consumer, environmental and trade union groups.[5] An Asia-Pacific Peoples' Environmental Network co-ordinates initiatives of some 300 groups in the region.
- The International Organization of Consumers Unions has been co-ordinating efforts of the world consumer movement through the Asia/Pacific Office in Penang, Malaysia.
- The Environment Liaison Centre International Office in Nairobi, Kenya has performed a similar function for the many environmental groups world-wide.
- The call for a world-wide moratorium on food irradiation has been endorsed by many other campaigning groups and regional conferences such as the Pesticide Action Network, the International Baby Food Action Network, the 1989 Regional Consumer Conference held in Japan and the Congress on Fate and Hope of the Earth held in Managua in 1989.

The International Food Irradiation Network (FIN) linking all concerned groups world-wide was established in 1989.

Links between developed and developing countries can be particularly valuable. Groups in Canada wrote to the Premier of Thailand to draw attention to the proposal to build a food irradiation plant in Bangkok. The plant was supplied by Canada using International Development Agency 'aid' money as a way of selling radioactive cobalt 60 and promoting Canadian irradiator technology. It was presented to the Thai food trade as a way of getting foods into the North American market. The reality was that the foods to be irradiated were not allowed to be sold in Canada. Thai groups wrote to the Canadian Premier at the same time causing considerable interest in the press in both countries.[6] The resulting public attention highlighted Canadian plans for irradiation as 'aid' to Jamaica and Malaysia.

Consumers

In the end, it is consumers who will decide whether or not to buy irradiated food. Consumer education is therefore vital. We believe that consumers have a right to all the facts about irradiated food — the uncertainties and the problems as well as the assurances. Unfortunately this is not what will be offered. The IAEA's pro-irradiation marketing strategy calls for promotional literature, slick videos, use of 'controlled communications' and hiring of public relations/issues management professionals to influence consumers to accept irradiation. It is vital that people who are concerned actively campaign to counter this propaganda.

What this book has shown is that consumers are being faced with a straight choice between good quality food and irradiation. There is no doubt which the consumer would prefer. The promoters will attempt to blur this question with reference to the effects of other preservation techniques and suggest that good quality food for all is an impossible dream. Implicit in this is the suggestion that, if the public wants unadulterated food, it must be prepared to pay more for it. In fact there is no reason why quality should always cost more. One clear example is given by a seafood company which had two production lines. One produced fish for Marks & Spencer and adhered to that company's exacting high standards. The other, serving different customers, had lower standards. The Marks & Spencer production line was cheaper to run.[7] Many companies outside of the food industry are learning the lesson — essentially, to find out what the consumer wants and supply it. The public wants quality as well as convenience, variety and availability. These are not conflicting but complementary demands. The companies that can satisfy them will succeed. Those that cannot will eventually fail — that is, provided consumers are clear in what they want and can withstand the onslaught of the advertising industry telling them what they want and so manipulating consumer demand.

The struggle to prevent irradiated food being foisted onto an unwilling public is, therefore, part of a wider struggle for food quality. This struggle is being fought not only within the framework of the official consumer organizations. These, as we have shown, are overwhelmingly opposed to irradiation. It is also being waged in every home, food store, supermarket, restaurant and canteen. It can be part of discussions within families and between friends at work and at social gatherings. This discussion of the issues is as important as more structured activities such as letter writing, raising issues within community organizations, or joining one of the groups around the country which is active on the issue.

This is what is happening all over the world. We hope that this book has given you some of the information you will need to become involved.

Checklist for action

1 Government/legislators

- In the UK, write to your Member of Parliament and to your Member of the European Parliament:

_________ MP, House of Commons, London SW1A 0AA

_________ MEP, 2 Queen Anne's Gate, London SW1

- Write to your elected representative at the local Council.

To find the name of your Councillor, MP or MEP, contact your local Council Information Dept or Citizens' Advice Bureau.

Outside the UK and Europe, substitute the relevant elected local/state/provincial/regional/federal/national elected representatives.

You or your group may wish to follow up the letters and meet with the elected representatives and discuss the actions you would like them to take.

Even in countries where irradiation is still banned, campaigning is needed. These countries will come under increasing pressure from other countries in the region. Also, the promoters of irradiation will not let the issue rest. If there is less to do locally and nationally because irradiation is banned, this may provide an opportunity to play a greater part in the international campaigns and help groups in developing countries.

2 The food industry

- Contact your local stores and supermarkets. Ask what their policy is on stocking irradiated food. Ask if they are prepared to give an undertaking that they will not (knowingly at least) sell irradiated food. Ask retailers to check with the companies which supply them that they will not be using irradiation on their products or buying irradiated ingredients. Produce your own positive list of companies and branded products that do not use irradiation. Share it with friends. Publicize it in local newsletters and through community organizations.
- Contact one of the organizations collaborating in the national and international positive list campaigns for the names, addresses and details of the current targets for letter writing. Let these organizations know the response you get from the food companies you write to.
- Remember to write to congratulate companies which take a responsible position as well as writing to let those which have not joined the lists know how you feel about the issue.

3 Non-governmental organizations

- Contact any organizations of which you are a member to find out where they stand on irradiation. Alerting the organization may prevent it from being used in the promotion of food irradiation. It will also help broaden the base of opposition. Remember to stress that opposition is needed now, whether or not people believe irradiation may have some benefits in the future.

Many organizations support the food irradiation campaign. Many of the activities above can be done on your own but most are more enjoyable and easier to sustain interest in if you do them as part of a group. You may wish to get a group of people together in your area or join an existing group which is involved in the campaign.

Contacts and addresses for some relevant organizations are given in Appendix D.

History of food irradiation

1916 • Sweden experiments with irradiation of strawberries
1921 • USA takes out irradiation patents
1930 • France takes out patents
1953 • Food irradiation announced as one of the 'atoms for peace' technologies
• US Army begins research into irradiation
1957 • West Germany irradiates spices
1958 • West Germany bans food irradiation
• USSR permits irradiation of potatoes
• USA classifies irradiation as a food additive; safety testing required
1960 • Canada permits irradiation of potatoes
1963 • USA permits irradiation of wheat, potatoes and bacon
1964 • IAEA, FAO and WHO set up Joint Expert Committee on Food Irradiation (JECFI) to review research data
• UK Dept of Health Report raises critical questions on the safety of irradiated food
1967 • UK prohibits irradiation and sale of imported irradiated food
1968 • US FDA withdraws permit for bacon when review finds adverse effects and poorly conducted experiments
• Accident at Hawaii irradiator. Contamination still detectable in 1980
1969 • First JECFI meeting, followed by report in 1970, on safety and wholesomeness of irradiated foods. Provisional clearance given for potatoes, wheat, and wheat products at specified doses
1971 • International Food Irradiation Project (IFIP) established with headquarters at Karlsruhe, West Germany

1972 • Japan grants permit to irradiate potatoes

1974 • Accidents at US irradiation facilities (Isomedix) (International Nutronics — uncovered in 1982)

1975 • Radiation leaks and worker exposure in accidents at irradiator facilities in USA (Radiation Technology) and Italy (Stimos)

1976 • Second JECFI meeting, followed by report in 1977, relaxes toxicology testing requirements for irradiated foods; clearances given (unconditionally) to potatoes, wheat, strawberries, papaya, chicken and (provisionally) to onions, rice, cod, redfish

• Australia grants first of two special permits to irradiate prawns to cover up bacterial contamination (second in 1979)

1977 • Worker exposed in accident at US irradiator facility (Radiation Technology)

1978 • Japanese baby food scandal — illegally irradiated vegetables used in baby foods since 1974. Studies show harmful effects in onions — onion irradiation not permitted

1980 • Third JECFI meeting, followed by report in 1981. Unconditional clearance given for all foods up to average dose 10 kGy

1981 • Fire at US irradiator (Becton Dickinson)

1982 • UK government Advisory Committee (ACINF) established

• Worker dies from irradiation accident in Norway

• IFIP research project terminated

1983 • UN Codex Alimentarius Commission approves International General Standards for Irradiated Foods and Recommended International Code of Practice for the Operation of Radiation Facilities for the Treatment of Foods

• Officials of IBT Ltd convicted of doing fraudulent research for government and industry — six years of food irradiation study data from 1970s, costing $4 million, is worthless

• Pollution Probe (Canada) publishes first critical report on food irradiation

• IAEA/FAO/WHO set up International Consultative Group on Food Irradiation (ICGFI) to promote food irradiation

1984 • US FDA proposal to permit unlabelled irradiated food withdrawn after public opposition

1985 • US Department of Energy (DOE) proposes six demonstration plants

• Campaigns launched by US Health and Energy Institute, National Coalition to Stop Food Irradiation, Food and Water, and others

• London Food Commission establishes working group on food irradiation

1986 • US FDA permits pork, fruit and vegetables (up to 1 kGy) and spices (up to 30 kGy)

• Abuse of irradiation uncovered. British food companies irradiate seafood in the Netherlands to hide bacterial contamination — known in the trade as 'Dutching'

• UK ACINF Committee reports no special safety problems with irradiated food up to 10 kGy. MPs question controls to prevent Dutching and conflict of interest of ACINF technical advisor

• IAEA Marketing Strategy outlines world-wide campaign for promoting food irradiation

• First meeting of Food Irradiation Network — groups opposing irradiation — held in the Netherlands

• Opinion poll in New Zealand/Aotearoa finds overwhelming opposition to irradiated food

• Denmark permits irradiated spices; revokes permit for potatoes

1987 • London Food Commission launches Food Irradiation Campaign to retain UK ban on irradiation with support of consumers, environmental groups, trade unions, public health, farmers and food industry representatives

• Public opinion polls find 93 per cent of UK population and 75 per cent of Canadians oppose food irradiation

• UK food manufacturers' body reverses position, calls for ban

• Leading UK supermarket chains declare opposition to selling irradiated food

• British Medical Association criticizes ACINF, suggests possible long-term health risks

• European Parliament votes against general clearance and for research on alternative preservation methods

• International Union of Foodworkers (IUF) calls for Europe-wide ban on irradiation

1987
- Further 'Dutching' scandal — Danish mussels
- US irradiated mushrooms scandal — Quaker Oats Co.
- European Scientific Committee endorses 1980 JECFI position
- International Organization of Consumers' Unions (IOCU) World Congress calls for moratorium until safety, control and other issues resolved. IOCU office in Penang, Malaysia, co-ordinates international consumer opposition
- Second European Food Irradiation Network meeting — Brussels
- Canadian Parliamentary report calls for safety research, recommends ban on irradiation of wheat
- Public demonstrations halt test marketing of irradiated potatoes in Canada, papayas in US and construction of irradiator in New Zealand/Aotearoa
- Australian Consumers Association report to Australian government endorses IOCU position
- California grocers' association opposes irradiated food
- US state of Maine bans sale of irradiated food

1988
- Second UK ACINF report recommends food irradiation but admits insufficient evidence on some safety issues; government retains ban because of concern over controls to prevent abuse
- Caesium leak at US irradiator (Radiation Sterilizers). State investigator calls for tighter regulation and controls
- New Zealand/Aotearoa government bans irradiation
- First meeting of Asia/Pacific Region Food Irradiation Network, Canberra, Australia
- Groups in Canada and Thailand unite to protest about Canadian aid money being used to fund Thailand irradiator. Similar 'aid' to Jamaica and Malaysia exposed
- IAEA 'research' meeting Bangkok (RPFI phase III) plans promotion of consumer acceptance in the Asia/Pacific region. Consumer groups alert President Cory Aquino about plans for promotion in the Philippines
- European Commission proposes directive to force all EEC countries to permit irradiation
- Australian parliamentary report recommends ban, calls on WHO to re-open safety investigation and produce a proper scientific report

- IOCU criticizes WHO promotion book, *Food Irradiation — a technique for preserving and improving the safety of food,* as biased, misleading and inaccurate. ICGFI produces promotion video, TV programme
- IAEA/WHO/FAO/ITC-UNCTAD/GATT conference in Geneva to promote food irradiation fails to achieve consensus, rejects statements that there are no unresolved safety problems. Of the 54 countries attending, 11 exprcss strong reservations over the final document
- International Food Irradiation Network (FIN) formally launched

1989
- Consumers force ICGFI to withdraw proposal to cripple existing labelling standards at Codex meeting, Ottawa, Canada
- 100 legislators support US federal bill to ban irradiation
- US Congress refuses further funds for DOE's irradiator programme
- New York and New Jersey ban sale of irradiated food. Ban Bills introduced in Vermont, New Hampshire, Massachusetts, Pennsylvania, Minnesota, Oregon, Washington and California
- Three workers exposed at (Delmed) irradiator in El Salvador
- Martin Welt, (ex-President Radiation Technology) convicted for conspiracy over safety violations. Forms new company, Alpha and Omega technology, to promote irradiation in SE Asia
- US FDA extends labelling requirement to 1990
- IUF Asia/Pacific Regional Congress opposes irradiation
- Public pressure stops plans for papaya irradiator in Hawaii, seafood irradiator in Alaska and apple irradiator in Washington
- Further 'Dutching' of seafood uncovered — companies under-irradiating to avoid port health detection and re-documenting consignments to avoid customs duty
- Canada extends permits for irradiation
- UK Retail Consortium, representing the major supermarket chains, joins consumer and public health/food law bodies in demanding the UK ban remain

1989 • European Parliament's Environment Committee votes for Europe-wide ban on irradiation (spices to be only exception)
• Danish Parliament's Consensus Conference rejects food irradiation as an inappropriate technology
• UK government announces intention to permit irradiation
• Conference on Fate and Hope of the Earth, Managua, Nicaragua, calls for world ban on irradiation
• Food Irradiation Network launches international seafood campaign

1990 • The tide turns?

Appendix A: Who uses irradiation, where and on what food?

A1 World list of irradiation plants

(Source: IAEA/FAO/AECL 1986)

COUNTRY	OPERATOR	LOCATION	* FOOD IRRADI-ATION	FOOD ITEM (TONS PER YEAR)	STARTING DATE
Argentina	CNEA (3 plants)	Buenos Aires	√	Spices, spinach, cocoa powder	1986
		Ezeiza			
		Mar del Plata			
Australia	Ansell International (2 plants)	Dandenong	√		
		Sydney			
	Johnson & Johnson	Sydney			
	Ansto (Lucas Heights)	Sydney	√	Research only	
		Brisbane	√		Planned
Bangladesh		Dhaka	√		Planned
		Chittagong	√	Potatoes, onions dried fish, pulses, frozen seafood, frog legs	1989
Belgium	IRE (2 plants)	Fleurus	√	Spices (350)	1981
		Fleurus	√	Dehydrated vegetables (700) deep-frozen foods (2,000)	
Brazil	Embrarad	Sao Paulo	√	Spices, dehydrated vegetables	1985
	IBRAS-CBO	Campinas-Est			
	Johnson & Johnson	San Jose Dos Campos			
Canada	Ethicon	Peterborough, Ontario			
	Isomedix	Whitby, Ontario			
	Sterirad	Markham, Ontario			
		Laval, Quebec	√	Spices	1987
Chile	CCEN	Santiago	√	Onions (200-300) potatoes (50-100) spices & dehydrated vegetables (20-30) chicken	1983

* Plants not marked √ are irradiating products other than food but might undertake food irradiation in the future

COUNTRY	OPERATOR	LOCATION	* FOOD IRRADI-ATION	FOOD ITEM (TONS PER YEAR)	STARTING DATE
China	Nuclear Research Institute	Shanghai	√	Potatoes, apples	1985
		Langzhou	√		1988
		Beijing	√		1988
		Jinan	√		1988
		Nanjing	√		1988
		Zhenzhou	√		1988
Czechoslovakia	Kova	Brno			
Cuba		Havana	√	Potatoes, onions, beans	1987
Denmark	Novo Alle	Copenhagen			
	Nunc	Roskilde			
	Molnlycke Steritex	Espergaerde			
	Risø		√	Spices	1986
Egypt	NCRT	Cairo			
El Salvador	Delka	San Salvador			
Federal Republic of Germany	Beiersdorf	Hamburg			
	Braun Melsungen	Melsungen			
	Ethicon	Hamburg			
	Gammaster Muenchen	Allershausen			
	Willy Ruesch	Rommelshausen			
Finland	Kolmi-Set Oy	Ilomantsi	√	Spices	1986
France	Conservatome (3 plants)	Dagneux (Lyon)	√	Spices (500-600)	1982
		Vannes	√	Poultry	1987
		Nice	√	Spices, vegetable seasonings	1986
	Caric	(Paris)	√	Spices, poultry (300)	1986
		Bretagne	√		Planned
		Marseille	√		Planned
		Orsay-Cedex	√		Planned
		Orleans	√		Planned
		Vende	√		Planned
German Democratic Republic	GAEC	Dresden			
	Cent. Int. Isotop. Radiat. Research	Weideroda	√	Onions (600)	1983
		Zwenkau		Garlic (4)	1986
	Queis Agric. Coop.	Spickendorf	√	Onions (4,000)	1986
	VEB Prowiko	Shoenebeck	√	Enzyme solution (300)	1986
Greece	Dimes	Athens			
Hungary	Medicor	Debrecen			
	Agroster	Budapest	√	Spices (200)	1982
		Budapest	√		Planned
India	IDEA	Bombay			
	Isomed	Bombay			
		Nasik			Planned
		Cochin			Planned
Indonesia		Pasar Jumat	√	Spices	1988
Iran	AEOI	Teheran			
Ireland	Becton Dickinson	Dublin			
	Imed	Letterkenny			
Israel	Sor-Van Radiation	Yavne	√	Spices (120)	1986
		Yavne	√		Planned
Italy	Ethicon	Rome			
	Gammarad	Bologna			
	Gammatom	Como			

COUNTRY	OPERATOR	LOCATION	* FOOD IRRADIATION	FOOD ITEM (TONS PER YEAR)	STARTING DATE
Italy (cont.)	ICO	Ascoli Picano			
	IPAS	Guanzate			
	Italgamma	Nettuno			
		Fucino	√		Planned
Ivory Coast (Cotes d'Ivoire)		Abidjan	√	Yam	1989
Jamaica			√		Planned
Japan	JAERI (2 plants)	Takasaki			
	Japan Radioisotope Association	Koka			
	Radia Industry	Takasaki			
	Terumo (2 plants)	Kofu			
	Tochigi-Siki	Tochigi			
	Shihoro Agricultural Cooperative	Hokkaido	√	Potatoes (8,000)	1973
Korea	KAERI (3 plants)	Seoul	√	Garlic powder	1985
		Seoul	√		Planned
Malaysia	Ansell International	Melaka			
	Nuclear Energy Unit (Puspati)	Bangi	√	Research/ demonstration	1988
		Kuala Lumpur			Planned
Mexico	ININ	Salazar			
	USDA	Tapachula			
		Mexico/USA border	√		Planned
Netherlands	Gammaster (2 plants)	Ede	√	Spices (100) Frozen products, poultry, spices, dehydrated vegetables, rice, dehydrated blood, egg powder, packaging materials	1978
	Pilot plant for Food Irradiation	Wageningen	√	Spices	1978
New Zealand	Coopers Animal Health	Upper Hutt			
Norway	Institute for Energy Technology	Kjeller	√	Spices (500)	1982
Pakistan		Lahore	√	Research	1987/8
Poland		Warsaw	√		Planned
		Poznan	√		Planned
		Przysucha	√		Planned
Saudi Arabia	Alshifa	Dammam			
	King Faisal Research Centre	Riyadh			
Singapore	Travenol Laboratories	Singapore			
South Africa	Atomic Energy/NDC	Pelindaba	√	Fruits, meats, onions, potatoes	1981
	Hepro	Tzaneen	√	Fruits, spices, onions, potatoes	1982
	Iso-Star	Johannesburg	√	Spices, dehydrated vegetables	1981
		Mulnerton	√	Fruit, spices, potatoes, onions, vegetables	1986

COUNTRY	OPERATOR	LOCATION	* FOOD IRRADI-ATION	FOOD ITEM (TONS PER YEAR)	STARTING DATE
South Africa (cont.)		Pretoria (3 plants)	√	Potatoes, onions, spices, meat, fish, chicken, fruit	1968-80
Sweden	Johnson & Johnson	Rotebro			
	Radona	Skaerhamn Island			
Switzerland	Steril Catgut	Neuhausen			
Taiwan	INER	Kneeling			
	ITRI	Taipei			
Thailand	Gammatron	Bangkok/ Paturntani	√	Shrimp, onions, garlic, sausage	1989
UK	Becton Dickinson	Plymouth			
	Ethicon	Edinburgh			
	Gillette	Reading			
	Isotron (4 plants)*	Bradford		Publicity/ demonstrations	
		Reading			
		Swindon (2 plants)	√		
	Isotron		√		Planned
	Johnson & Johnson	Slough			
	Surgikos	Livingston			
	Swann Morton	Sheffield			
USA	American Pharmaseal (Convertors Operation) (2 plants)	El Paso, TX			
	Applied Radiant Energy	Lynchburg, VA	√		1986
	Becton Dickinson (3 plants)	North Canaan, CT			
		Broken Bow, NB			
		Sumter, SC			
	Cobe Laboratories	Denver, CO			
	Ethicon (2 plants)	Somerville, NJ			
		San Angelo, TX			
	International Nutronics (2 plants)	Irvine, CA			
		Palo Alto, CA			
	Isomedix (10 plants)	Libertyville, IL	√	Spices (500)	1984
		Morton Grove, IL	√		
		Northborough, MA	√		
		Columbus, MS	√		
		Parsippany, NJ			
		Whippany, NJ			
		Groveport, OH			
		Vega Alta, Puerto Rico	√		
		Spartanburg, SC	√		
		Sandy City, UT	√		
	Johnson & Johnson	Sherman, TX			
	Neutron Products	Dickerson, MD			
	Precision Materials	Mine Hill, NJ			
	Radiation Sterilizers (4 plants)	Tustin, CA	√	Spices (500)	1984
		Decatur, GA	√		
		Schaumberg, IL	√		
		Fort Worth, TX	√		
		Westerville, OH			

*Isotron plants could irradiate food when the process is legalized

COUNTRY	OPERATOR	LOCATION	* FOOD IRRADIATION	FOOD ITEM (TONS PER YEAR)	STARTING DATE
	Radiation Technology (4 plants)	West Memphis, AR	√	Spices (500)	1984
		Haw River, NC	√		
		Salam, NJ			
		Rockaway, NJ	√		
USA (cont.)	Sherwood Medical (3 plants)	Norfolk, NB			
		Deland, FL			
		Commerce, TX			
	Surgikos	Arlington, TX			
	Travenol Laboratories	Aibinito, Puerto Rico			
	3M Company	Brookings, SD			
	US DOE	Washington	√		Planned
		Iowa	√		Planned
		Alaska	√		Planned
		Oklahoma	√		Planned
		Hawaii	√		Planned
		Florida	√		Planned
USSR (1981)	Grand Total: 11 industrial gamma irradiators (4 medical, 3 food, 4 chemical)		√√√	Potatoes, onions, cereals, fruits, vegetables, meat, poultry	1960
	Odessa Port Elevator RDU	Odessa	√	Grains (400,000)	1983
Venezuela	IVIC	Caracas			
Vietnam		Hanoi	√		Planned
Yugoslavia	Nuclear Institute	Vinca			
		Belgrade	√	Spices	
	Ruder Boskovic Institute	Zagreb	√	Black pepper	1985

A2: List of countries that have cleared irradiated food for human consumption

Sources: WHO 1988, IAEA 1988, Webb & Lang

Country	Product	Purpose of irradiation	Type of clearance	Dose permitted (kGy)	Date approved
Argentina	strawberries	shelf-life extension	unconditional	2.5 max.	1987
	potatoes	sprout inhibition	unconditional	0.03 to 0.15	1987
	onions	sprout inhibition	unconditional	0.02 to 0.15	1987
	garlic	sprout inhibition	unconditional	0.02 to 0.15	1987
Australia	shrimp	decontamination	special permits*		1976 & 1979
Bangladesh	chicken	shelf-life extension/decontamination	unconditional	up to 8	1983
	papaya	insect disinfestation/control of ripening	unconditional	up to 1	1983
	potatoes	sprout inhibition	unconditional	up to 0.15	1983
	wheat and ground wheat products	insect disinfestation	unconditional	up to 1	1983
	fish	shelf-like extension/decontamination/insect disinfestation	unconditional	up to 2.2	1983
	onions	sprout inhibition	unconditional	up to 0.15	1983
	rice	insect disinfestation	unconditional	up to 1	1983
	frog legs	decontamination	provisional		
	shrimp	shelf-life extension/decontamination	provisional		
	mangoes	shelf-life extension/insect disinfestation/control ripening	unconditional	up to 1	1983
	pulses	insect disinfestation	unconditional	up to 1	1983
	spices	decontamination/insect disinfestation	unconditional	up to 10	1983
Belgium	potatoes	sprout inhibition	provisional	up to 0.15	1980
	strawberries	shelf-life extension	provisional	up to 3	1980
	onions	sprout inhibition	provisional	up to 0.15	1980
	garlic	sprout inhibition	provisional	up to 0.15	1980
	shallots	sprout inhibition	provisional	up to 0.15	1980
	black/white pepper	decontamination	provisional	up to 10	1980
	paprika powder	decontamination	provisional	up to 10	1980
	arabic gum	decontamination	provisional	up to 10	1983
	spices (78 different products)	decontamination	provisional	up to 10	1983

* Process remains banned

Country	Product	Purpose of irradiation	Type of clearance	Dose permitted (kGy)	Date approved
Belgium cont.	(semi)-dried vegetables (7 different products)	decontamination	provisional	up to 10	1983
Brazil	rice	insect disinfestation	unconditional	up to 1	1985
	potatoes	sprout inhibition	unconditional	up to 0.15	1985
	onions	sprout inhibition	unconditional	up to 0.15	1985
	beans	insect disinfestation	unconditional	up to 1	1985
	maize	insect disinfestation	unconditional	up to 0.5	1985
	wheat	insect disinfestation	unconditional	up to 1	1985
	wheat flour	insect disinfestation	unconditional	up to 1	1985
	spices (13 different products)	decontamination/insect disinfestation	unconditional	up to 10	1985
	papaya	insect disinfestation/control of ripening	unconditional	up to 1	1985
	strawberries	shelf-life extension	unconditional	up to 3	1985
	fish and fish products (fillets, salted, smoked dried, dehydrated)	shelf-life extension/decontamination/insect disinfestation	unconditional	up to 2.2	1985
	poultry	shelf-life extension/decontamination	unconditional	up to 7	1985
Bulgaria	potatoes	sprout inhibition	experimental batches	0.1	1972
	onions	sprout inhibition	experimental batches	0.1	1972
	garlic	sprout inhibition	experimental batches	0.1	1972
	grain	insect disinfestation	experimental batches	0.3	1972
	dry food concentrates	insect disinfestation	experimental batches	1	1972
	dried fruits	insect disinfestation	experimental batches	1	1972
	fresh fruits (tomatoes, peaches, apricots cherries, raspberries, grapes)	shelf-life extension	experimental batches	2.5	1972
Canada	potatoes	sprout inhibition	unconditional	up to 0.1	1960
	onions	sprout inhibition	unconditional	up to 0.15	1965
	wheat, flour, wholemeat	insect disinfestation	unconditional	up to 0.75	1969
	poultry	decontamination	test marketing	up to 7	1973

Country	Product	Purpose of irradiation	Type of clearance	Dose permitted (kGy)	Date approved
Canada *cont.*	cod and haddock fillets	shelf-life extension	test marketing	up to 1.5	1973
	spices and certain dried vegetables' seasonings	decontamination	unconditional	up to 10	1984
	onion powder	decontamination	unconditional	up to 10	1983
Chile	potatoes	sprout inhibition	experimental batches		1974
			test marketing		1982
			unconditional	up to 0.15	
	papaya	insect disinfestation	unconditional	up to 1	1982
	wheat and ground wheat products	insect disinfestation	unconditional	up to 1	1982
	strawberries	shelf-life extension	unconditional	up to 3	1982
	chicken	decontamination	unconditional	up to 7	1982
	onions	sprout inhibition	unconditional	up to 0.15	1982
	rice	insect disinfestation	unconditional	up to 1	1982
	fish and fish products	shelf-life extension/decontamination/insect disinfestation	unconditional	up to 2.2	1982
	cocoa beans	decontamination/ insect disinfestation	unconditional	up to 5	1982
	dates	insect disinfestation	unconditional	up to 1	1982
	mangoes	shelf-life extension/insect disinfestation/control of ripening	unconditional	up to 1	1982
	pulses	insect disinfestation	unconditional	up to 1	1982
	spices and condiments	decontamination/insect disinfestation	unconditional	up to 10	1982
China	potatoes	sprout inhibition	unconditional	up to 0.20	1984
	onions	sprout inhibition	unconditional	up to 0.15	1984
	garlic	sprout inhibition	unconditional	up to 0.10	1984
	peanuts	insect disinfestation	unconditional	up to 0.40	1984
	grain	insect disinfestation	unconditional	up to 0.45	1984
	mushrooms	growth inhibition	unconditional	up to 1	1984
	sausage	decontamination	unconditional	up to 8	1984
Czechoslovakia	potatoes	sprout inhibition	experimental batches	up to 0.1	1976
	onions	sprout inhibition	experimental batches	up in 0.08	1976

Country	Product	Purpose of irradiation	Type of clearance	Dose permitted (kGy)	Date approved
Czech. cont.	mushrooms	growth inhibition	experimental	up to 2	1976
Denmark	potatoes	sprout inhibition	unconditional *	up to 0.15	1965
	spices and herbs	decontamination	unconditional	up to 10 average up to 15 max.	1985
Finland	dry and dehydrated spices and herbs	decontamination	unconditional	up to 10 average	1987
	all foods for patients requiring a sterile diet	sterilization	unconditional	unlimited	1987
France	potatoes	sprout inhibition	provisional	0.075-0.15	1972
	onions	sprout inhibition	provisional	0.075-0.15	1977
	garlic	sprout inhibition	provisional	0.075-0.15	1977
	shallots	sprout inhibition	provisional	0.075-0.15	1977
	spices and aromatic substances (72 products including powdered onion and garlic)	decontamination	unconditional	up to 11	1983
	gum arabic	decontamination	unconditional	up to 9	1985
	muesli-like cereal	decontamination	unconditional	up to 10	1985
	dehydrated vegetables	decontamination	unconditional	up to 10	1985
	mechanically deboned poultry meat	decontamination	unconditional	up to 5	1985
	dried fruits	insect disinfestation	unconditional	1 max.	1988
	dried vegetables	insect disinfestation	unconditional	1 max.	1988
German Democratic Republic	onions	sprout inhibition	test marketing	50	1981
	onions	sprout inhibition	unconditional	20	1984
	enzyme solutions	decontamination	unconditional	10	1983
	spices	decontamination	provisional	up to 10	1982

* permit revoked as unnecessary 1988

Country	Product	Purpose of irradiation	Type of clearance	Dose permitted (kGy)	Date approved
Hungary	potatoes	sprout inhibition	test marketing	0.1	1969
	potatoes	sprout inhibition	test marketing	0.15 max.	1972
	potatoes	sprout inhibition	test marketing	0.15 max.	1973
	onions	sprout inhibition	test marketing		1973
	strawberries	shelf-life extension	test marketing		1973
	mixed spices (black pepper, cumin, paprika, dried garlic: for use in sausages)	decontamination	experimental batches	5	1974
	onions	sprout inhibition	test marketing	0.06	1975
	onions	sprout inhibition	experimental batches	0.06	1976
	mixed dry ingredients for canned hashed meat	decontamination	experimental batches	5	1976
	potatoes	sprout inhibition	test marketing	0.1	1980
	onions	sprout inhibition	experimental patches	0.05	1980
	onions (for dehydrated flakes processing)	sprout inhibition	test marketing	0.05	1980
	mushrooms (*Agaricus*)	growth inhibition	test marketing	2.5	1981
	strawberries	shelf-life extension	test marketing	2.5	1981
	potatoes	sprout inhibition	test marketing	0.1	1981
	potatoes	sprout inhibition	test marketing	0.1	1981
	spices for sausage production	decontamination	test marketing	5	1982
	strawberries	shelf-life extension	test marketing	2.5	1982
	mushrooms (*Agaricus*)	growth inhibition	test marketing	2.5	1982
	mushrooms (*Pleurotus*)	growth inhibition	test marketing	3	1982
	grapes	shelf-life extension	test marketing	2.5	1982
	cherries	shelf-life extension	test marketing	2.5	1982
	sour cherries	shelf-life extension	test marketing	2.5	1982
	red currants	shelf-life extension	test marketing	2.5	1982
	onions	sprout inhibition	unconditional	0.05-0.07	1982
	spices for sausage	decontamination	test marketing	5	1982
	pears	shelf-life extension	test marketing	2.5	1982
	pears	shelf-life extension	test marketing	1.0+$CaCl_2$ treatment	1983

Country	Product	Purpose of irradiation	Type of clearance	Dose permitted (kGy)	Date approved
Hungary cont.	spices	decontamination	test marketing	5	1983
	potatoes (for processing into flakes)	sprout inhibition	test marketing	0.1	1983
	frozen chicken	decontamination	test marketing	4	1983
	sour cherries (canned)		conditional	0.2 average	1984
					1985
	black pepper	decontamination	conditional	6 minimum	1985
	spices	decontamination	conditional	5-6 minimum	1986
	spices	decontamination	unconditional	8, 6 average	1986
India	potatoes	sprout inhibition	unconditional	Codex Standard	1986
	onions	sprout inhibition	unconditional	Codex Standard	1986
	spices	disinfection	for export only	Codex Standard	1986
	frozen shrimps and frog legs	disinfection	for export only	Codex Standard	1986
Indonesia	dried spices	decontamination	unconditional	10 max.	1987
	tuber and root crops (potatoes, shallots, garlic and rhizomes)	sprout inhibition	unconditional	0.15 max.	1987
	cereals	disinfestation	unconditional	1 max.	1987
Israel	potatoes	sprout inhibition	unconditional	0.15 max.	1967
	onions	sprout inhibition	unconditional	0.10 max.	1968
	poultry and poultry sections	shelf-life extension/decontamination	unconditional	7 max.	1982
	onions	sprout inhibition	unconditional	0.15	
	garlic	sprout inhibition	unconditional	0.15	1985
	shallots	sprout inhibition	unconditional	0.15	1985
	spices (36 different products)	decontamination	unconditional	10	1985
	fresh fruits and vegetables	disinfestation	unconditional	1 average	1987
	grains, cereals, pulses, cocoa & coffee beans, nuts, edible seeds	disinfestation	unconditional	1 average	1987

Country	Product	Purpose of irradiation	Type of clearance	Dose permitted (kGy)	Date approved
Israel cont.	mushrooms, strawberries	shelf-life extension	unconditional	3 average	1987
	poultry and poultry sections	decontamination	unconditional	7 average	1987
	spices & condiments dehydrated & dried vegetables, edible herbs	decontamination	unconditional	10 average	1987
	poultry feeds	decontamination	unconditional	15 average	1987
Italy	potatoes	sprout inhibition	unconditional	0.075-0.15	1973
	onions	sprout inhibition	unconditional	0.075-0.15	1973
	garlic	sprout inhibition	unconditional	0.075-0.15	1973
Japan	potatoes	sprout inhibition	unconditional	0.15 max.	1972
Mexico	onions	(details not included in IAEA/WHO publications)			
	fresh fruit				
	spices				
	cocoa				
	nuts				
Netherlands	asparagus	shelf-life extension/growth inhibition	experimental batches	2 max.	1969
	cocoa beans	insect disinfestation	experimental batches	0.7 max.	1969
	strawberries	shelf-life extension	experimental batches	2.5 max.	1969
	mushrooms	growth inhibition	unconditional	2.5 max.	1969
	deep frozen meals	sterilization	hospital patients	25 min.	1969
	potatoes	sprout inhibition	unconditional	0.15 max.	1970
	shrimps	shelf-life extension	experimental batches	0.5-1	1970
	onions	sprout inhibition	experimental batches	0.15	1971
	spices and condiments	decontamination	experimental batches	8-10	1971
	poultry, eviscerated (in plastic bags)	shelf-life extension	experimental batches	3 max.	1971
	chicken	shelf-life extension/decontamination	unconditional	3 max.	1976
	fresh, tinned and liquid foodstuffs	sterilization	hospital patients	25 min.	1972
	spices	decontamination	provisional	10	1974

Country	Product	Purpose of irradiation	Type of clearance	Dose permitted (kGy)	Date approved
Netherlands cont.	powdered batter mix	decontamination	test marketing	1.5	1974
	vegetable filling	decontamination	test marketing	0.75	1974
	endive (prepared, cut)	shelf-life extension	test marketing	1	1975
	onions	sprout inhibition	unconditional	0.05 max.	1975
	spices	decontamination	provisional	10	1975
	peeled potatoes	shelf-life extension	test marketing	0.5	1976
	chicken	shelf-life extension/decontamination	unconditional	3 max.	1976
	shrimps	shelf-life extension	test marketing	1	1976
	fillets of haddock, coal-fish whiting	shelf-life extension	test marketing	1	1976
	fillets of cod and plaice	shelf-life extension	test marketing	1	1976
	fresh vegetables (prepared, cut, soup greens)	shelf-life extension	test marketing	1	1977
	spices	decontamination	provisional	10	1978
	frozen frog legs	decontamination	provisional	5	1978
	rice and ground rice products	insect disinfestation	provisional	1	1979
	rye bread	shelf-life extension	provisional	5 max.	1980
	spices	decontamination	provisional	7 max.	1980
	frozen shrimp	decontamination	provisional	7 max.	1980
	malt	decontamination	provisional	10 max.	1983
	boiled and cooled shrimp	shelf-life extension	provisional	1 max.	1983
	frozen shrimp	decontamination	provisional	7 max.	1983
	frozen fish	decontamination	provisional	6 max.	1983
	egg powder	decontamination	provisional	6 max.	1983
	dry blood protein	decontamination	provisional	7 max.	1983
	dehydrate vegetables	decontamination	provisional	10 max.	1983
	refrigerated snacks of minced meat	shelf-life extension	test marketing	2	1984
New Zealand	herbs and spices (one batch)	decontamination	provisional*	8	1985

* process formally baned 1988

Country	Product	Purpose of irradiation	Type of clearance	Dose permitted (kGy)	Date approved
Norway	spices	decontamination	unconditional	up to 10	
Philippines	potatoes	sprout inhibition	provisional	0.15 max.	1972
	onions	sprout inhibition	provisional	0.07	1981
	garlic	sprout inhibition	provisional	0.07	1981
	onions and garlic	sprout inhibition	test marketing		1984
			test marketing		1986
Poland	potatoes	sprout inhibition	provisional	up to 0.15	1982
	onions	sprout inhibition	provisional		1983
Republic of Korea	potatoes	sprout inhibition	unconditional	0.15 max.	1987
	onions	sprout inhibition	unconditional	0.15 max.	1987
	garlic	sprout inhibition	unconditional	0.15 max.	1987
	chestnuts	sprout inhibition	unconditional	0.25 max.	1987
	fresh and dried mushrooms	growth inhibition/insect disinfestation	unconditional	1 max.	1987
South Africa	potatoes	sprout inhibition	unconditional	0.12-0.24	1977
	dried bananas	insect disinfestation	provisional	0.5 max.	1977
	avocados	insect disinfestation	provisional	0.1 max.	1977
	onions	sprout inhibition	unconditional	0.05-0.15	1978
	garlic	sprout inhibition	unconditional	0.1-0.20	1978
	chicken	shelf-life extension/decontamination	unconditional	2-7	1978
	papaya	shelf-life extension	unconditional	0.5-1.5	1978
	mango	shelf-life extension	unconditional	0.5-1.5	1978
	strawberries	shelf-life extension	unconditional	1-4	1978
	bananas	shelf-life extension	unconditional		1982
	litchis	shelf-life extension	unconditional		1982
	pickled mango (achar)	shelf-life extension	unconditional		1982

Country	Product	Purpose of irradiation	Type of clearance	Dose permitted (kGy)	Date approved
South Africa cont.	avocados	shelf-life extension	unconditional		1982
	frozen fruit juices	shelf-life extension	unconditional		
	green beans		unconditional		
	tomatoes	control of ripening	unconditional		
	brinjals		unconditional		
	soya pickle products		unconditional		
	ginger		unconditional	(details not included in IAEA/WHO publications)	
	vegetable paste		unconditional		
	bananas (dried)	insect disinfestation	unconditional		
	almonds	insect disinfestation	unconditional		
	cheese powder	insect disinfestation	unconditional		
	yeast powder		unconditional		
	herbal tea		unconditional		
	various spices		unconditional		
	various dehydrated vegetables		unconditional		
Spain	potatoes	sprout inhibition	unconditional	0.05-0.15	1969
	onions	sprout inhibition	unconditional	0.08 max.	1971
Thailand	onions	sprout inhibition	unconditional	0.1 max.	1973
	potatoes, onions, garlic	sprout inhibition	unconditional	0.15	1986
	dates	disinfestation	unconditional	1	1986
	mangoes, papaya	disinfestation/delay of ripening	unconditional	1	1986
	wheat, rice, pulses	disinfestation	unconditional	1	1986
	cocoa beans	disinfestation	unconditional	1	1986
	fish and fishery products	disinfestation	unconditional	1	1986
	fish and fishery products	reduce microbial load	unconditional	2.2	1986
	strawberries	shelf-life extension	unconditional	3	1986
	nam	decontamination	unconditional	4	1986
	moo yor	decontamination	unconditional	5	1986
	sausage	decontamination	unconditional	5	1986
	frozen shrimps	decontamination	unconditional	5	1986

Country	Product	Purpose of irradiation	Type of clearance	Dose permitted (kGy)	Date approved
Thailand cont.	cocoa beans	reduce microbial load	unconditional	5	1986
	chicken	decontamination/shelf-life extension	unconditional	7	1986
	dehydrated spices & condiments	insect disinfestation	unconditional	1	1986
	onions and onion powder	decontamination	unconditional	10	1986
USSR	potatoes	sprout inhibition	unconditional	0.1 max	1958
	potatoes	sprout inhibition	unconditional	0.3 (1 MeV-electrons)	1973
	grain	insect disinfestation	unconditional	0.3	1959
	fresh fruits and vegetables	shelf-life extension	experimental batches	2-4	1964
	semi-prepared raw beef, pork & rabbit products (in plastic bags)	shelf-life extension	experimental batches	6-8	1964
	dried fruits	insect disinfestation	unconditional	1	1966
	dry food concentrates (buckwheat, mush, gruel, rice, pudding)	insect disinfestation	unconditional	0.7	1966
	poultry, eviscerated (in plastic bags)	shelf-life extension	experimental batches	6	1966
	culinary prepared meat products (fried meat, entrecote) (in plastic bags)	shelf-life extension	test marketing	8	1967
	onions	sprout inhibition	testmarketing	0.06	1967
	onions	sprout inhibition	unconditional	0.06	1973
UK	any food for consumption by patients who require a sterile diet as an essential factor in their treatment	sterilization	hospital patients		1969

Country	Product	Purpose of irradiation	Type of clearance	Dose permitted (kGy)	Date approved
USA	can-packed bacon		*		1963
	wheat and wheat flour	insect disinfestation	unconditional	0.2-0.5	1963
	white potatoes	shelf-life extension	unconditional	0.05-0.1	1964
	white potatoes	shelf-life extension	unconditional	0.05-0.15	1965
	spices and dry vegetable seasonings (38 commodities)	decontamination/insect disinfestation	unconditional (revised 1986)	30 max.	1983
	dry or dehydrated enzyme preparations (including immobilized enzyme preparations)	control of insects and/or micro organisms	unconditional	10 kGy max.	1985
	pork carcasses or fresh, non-heat processed cuts of pork carcasses	control of *Trichinella spiralis*	unconditional	0.3 min.– 1.0 max.	1985
	fresh fruits and vegetables	delay of maturation and insect disinfestation	unconditional	1	1986
	dry or dehydrated enzyme preparations	decontamination	unconditional	10	1986
	herbs, spice blends and vegetable seasonings.	decontamination	unconditional	30	1986
Uruguay	potatoes	sprout inhibition	unconditional		1970
Yugoslavia	cereals	insect disinfestation	unconditional	up to 10	1984
	legumes	insect disinfestation	unconditional	up to 10	1984
	onions	sprout inhibition	unconditional	up to 10	1984
	garlic	sprout inhibition	unconditional	up to 10	1984
	potatoes	sprout inhibition	unconditional	up to 10	1984
	dehydrated fruits & vegetables	sprout inhibition	unconditional	up to 10	1984
	dried mushrooms		unconditional	up to 10	1984
	egg powder	decontamination	unconditional	up to 10	1984

* permit revoked on safety grounds 1968

Country	Product	Purpose of irradiation	Type of clearance	Dose permitted (kGy)	Date approved
Yugoslavia cont.	herbals teas, tea extracts	decontamination	unconditional	up to 10	1984
	fresh poultry	shelf-life extension/decontamination	unconditional	up to 10	1984

Recommendations published by international organizations

The following recommendations have been published by the IAEA/FAO/WHO joint expert committee (JECFI)

Date of recommendation	Product	Purpose of irradiation	Type of clearance	Dose permitted (kGy)
1969	potatoes	sprout inhibition	provisional	0.15 max.
	wheat and ground wheat products	insect disinfestation	provisional	0.75 max.
1976	potatoes	sprout inhibition	unconditional	0.03-0.15
	onions	sprout inhibition	provisional	0.02-0.15
	papaya	insect disinfestation	unconditional	0.5-1
	strawberries	shelf-life extension	unconditional	1-3
	wheat and ground wheat products	insect disinfestation	unconditional	0.15-1
	rice	insect disinfestation	provisional	0.1-1
	chicken	shelf-life extension/decontamination	unconditional	2-7
	cod & redfish	shelf-life extension/decontamination	provisional	2-2.2
1980	any food product	sprout inhibition/shelf-life extension/decontamination insect disinfestation/control of ripening/growth inhibition	unconditional	up to 10

Appendix B: The food irradiation scandal

B1 Irradiated food: conflict of interest

EDM 713, House of Commons Order Paper. 9 April 1986

Mr Dennis Skinner
Mr Richard Caborn
Mr Robert Litherland
Mr Tony Lloyd
Mr Brian Sedgemore
Mr Ernie Ross

That this House expresses concern about conflict of interest between the work of Her Majesty's Government's Advisory Committee on Irradiated and Novel Foods and the changing position of companies who stand to benefit from its recommendations; notes that the predictions of the main recommendation of the Committee, that the current ban on irradiating food can be lifted, have been widely leaked, not least by Mr Frank Ley, the Committee's technical and economic adviser; notes that Mr Ley is the Marketing Director of ISOTRON, a company which is in a semi-monopolistic position to take advantage of a change in law allowing food to be irradiated in Britain; notes that while the Committee sat, ISOTRON, despite its existing production overcapacity, commissioned a new plant and raised capital through a flotation on the Stock Exchange; notes the rise in ISOTRON's capital value when the financial press linked the future of the company to the impending recommendations of the Advisory Committee; believes that Mr Ley's high public profile, designed to help ISOTRON financially, was incompatible with his rôle as regards the Advisory Committee; calls on all the directors of ISOTRON to give an account of their share interests and charges on them since flotation; and calls on Her Majesty's Government to make a statement and the Stock Exchange to carry out an investigation, so as to satisfy themselves that nothing improper has occurred.

B2 Irradiated food: illegal importation

EDM 714 House of Commons Order Paper. 9 April 1986

Mr Frank Cook
Mr Robert Litherland
Mr Brian Sedgemore
Mr Richard Caborn
Mr Ernie Ross
Mr Tony Lloyd

That this House notes with grave concern the admission of a company from the Imperial Food Group that, having had a consignment of prawns condemned by the Southampton Port Health Authority on the grounds of bacterial contamination, it had the consignment irradiated by Gammaster BV in the Netherlands to conceal such bacterial contamination and caused them to be reimported; notes that this practice is illegal under the United Kingdom Food (Control of Irradiation) Regulations; notes too that it is also contrary to the recommendations of the Joint Export Committee of the World Health Organisation Food and Agriculture Organisation of the United Nations and the International Atomic Energy Agency, which bodies demand that irradiation be used solely to extend the shelf life of food otherwise wholesome and never be used to conceal contamination, thereby rendering saleable food that is not truly wholesome; notes further that this instance highlights the inadequacy of existing food monitoring provisions and standards of enforcement procedures, presenting a potential public health hazard; calls on the Imperial Food Group to give to their shareholders, and to the shareholders of United Biscuits, a full explanation of how and why they became involved in such a scandalous practice; requires them to give, too, a public undertaking that such a major misdemeanour will not be repeated; calls upon Her Majesty's Government to alert all port health and trading standards authorities to the possible prospect of such malpractice: and finally urges the Director of Public Prosecutions to investigate these matters fully with a view to initiating appropriate legal proceedings against the offending company.

CERTIFICATE OF GAMMA IRRADIATION No. 214

This is to certify that :

GAMMASTER B.V. - Ede - Holland

has given an irradiation treatment
to the following goods :

CUSTOMER : Young's Seafoods Ltd., London,

PRODUCT : Bulk IQF cooked & peeled prawns 300/400

QUANTITY : 1. 1278 x 38 lbs
2. 180 x 38 lbs and 572 x 25 lbs

CHARGENUMBER : 1. 850423
2. 850423-24/NW5-029

IRRADIATION DATE : 1. 23-04-1985
2. 23/24-04-1985

IRRADIATION MODE : JS 9000 IRRADIATOR

IRRADIATION DOSE : 3 kGy

GAMMASTER B.V.
Ede - Holland

Acc. Controller :

Acc. Plant Manager :

Daily Mail, Monday, March 3, 1986

We broke law with gamma ray prawns says food firm

By STEPHEN LEATHER, Consumer Affairs Correspondent

ONE of Britain's biggest food groups has sold radiation-treated prawns in contravention of health regulations.

The Imperial Group admits it broke the law last year after a consignment of prawns from Malaysia failed quality-control tests.

Instead of scrapping the seafood, it was shipped to Holland, sterilised with gamma radiation and then brought back and sold under the Admiral label to caterers.

Imperial Group — its products include Courage beer, Ross frozen foods, Golden Wonder and Young's Seafoods — said : 'In January 1985 we bought two containers of Malaysian warm water prawns. They were imported through Southampton, where they were tested by the health authorities. Then they went into public cold store.

'In February we took them out. We didn't have to test them again, but we did. We tested 46 batches. Twenty passed but 26 did not — our standards for these tests are very nigh.

'There was nothing to stop us selling the prawns because they had been passed by the health authorities — but our standards are higher. The decision was taken to send them to Holland for irradiation. Holland is a world leade: in this type of treatment.'

A certificate shows the Dutch firm of Gammaster treated three batches of cooked, peeled prawns in April. They were then returned to Britain.

The spokesman said that by the end of May, all the radiation-treated prawns had been sold, mainly to Indian restaurants and caterers.

The Department of Health confirmed that it is illegal to sell irradiated food in Britain and that the trading standards office at the port of entry would be investigating.

Prosecution could result in a fine of up to £2,000 or three months' jail.

Irradiation is widely used to sterilise food in Europe and America — there is no evidence that treated food is unsafe. The treatment allows it to be kept for much longer, which can reduce its nutritional value.

Ministry of Agriculture, Fisheries and Food
From the Parliamentary Secretary

Dear Mr Lang,

Thank you for your letter of 9 April enclosing a copy of a certificate of irradiation for a consignment of imported prawns.

Enforcement of the Food (Control of Irradiation) Regulations is a matter for local and port health authorities. I know that officials of the Department of Health have discussed this incident with the Dover Port Health Authority, who are investigating and will I am sure take action if they think it appropriate. A single incident of this nature certainly cannot be taken as evidence of widespread evasion of the law and I have no knowledge of any other incidents, which in any case would be a matter for local authorities. Importers are required to comply with our domestic legislation regardless of the country of origin and, if action is warranted, it will be taken by the appropriate authorities here.

I have copied your letter and enclosure to the Department of Health and Social Security.

PEGGY FENNER

B3 Illegal food irradiation

EDM 950, House of Commons Order Paper. 9 June 1989

Mr Frank Cook
Mr Eric Martlew
Hilary Armstrong
Mr Martyn Jones
Mr Nigel Griffiths
Marjorie Mowlam

That this House notes that since learning in 1986 of the covert and illegal use of food irradiation, by such companies as Youngs Seafoods of the Imperial Food Group and the Flying Goose Company of Allied Lyons, purposely to conceal bacterial contamination of shellfish that would otherwise render them unsaleable, and on which references Her Majesty's Government refused to act, further disturbing and continuing reports have come to light including the irradiation by Gammaster BV in the Netherlands of contaminated mussels previously rejected by Danish authorities for subsequent sale back in Denmark and the similar treatment of components for baby foods labelled as animal feed in Japan: urges government investigation of Gammaster, who appear willing to supply British companies dealing illegally in such commodities, of Allways Transport, who appear able and willing to import them illegally and distribute them at will throughout the United Kingdom, of Hank DeBruijne, who appear to specialise in arranging treatment of reject consignments and apply a technique of under-irradiation or partial dose to leave some bacteria active and so confuse port health authorities who might otherwise be suspicious of an abnormally low bacterial load, and of United Kingdom importers, Smith Fish of Grimsby, S.P.I. of Birmingham and Lyons Seafoods of Warminster who appear to trade covertly and consistently with the Dutch companies offering these questionable services; and demands official action to expose further such nefarious practices and effective measures to put a stop to such abuse.

B4 A newspaper investigation reveals evidence of 'Dutching'

THE SUNDAY TIMES 6 AUGUST 1989

Illegal trade in spoiled seafood uncovered

by Barrie Penrose

AN illegal trade in seafood which is irradiated to kill traces of salmonella before being brought to Britain has been exposed by The Sunday Times.

The traffic in the food came to light after reporters, posing as buyers, were offered a 16-ton consignment of Indian prawns which had been seized and condemned by American health officials because of salmonella contamination.

Last night the 815 cases of raw prawns were in a dockside warehouse in New Jersey awaiting another buyer.

They are being offered for sale by Landauer, a London firm of commodity brokers which claims to have bought the seafood from the original American importer.

Landauer sent out a telex last week telling would-be British clients: "There is an Indian pd [peeled and deveined] lot rejected in USA for salmonella . . . if interested, pls bid."

One of Landauer's dealers offered to send the prawns "on holiday to Holland", a trade expression for diverting them to an irradiation plant in The Netherlands, before they come to Britain.

Health officials in a British port testing for salmonella would find the prawns "clean", and be unable to detect that they had been irradiated.

Importing irradiated food was banned in Britain in 1967. Irradiation kills most of the bacteria and hides the fact that salmonella has been present in food.

On Wednesday, Stephen Brown, a dealer at Landauer's offices in the City, talked about "Dutching" to a Sunday Times reporter posing as a potential buyer.

Asked if the prawns would pass health tests at a British port, Brown said: "Well, they won't if they came into England directly. But if they went into Holland and Belgium, yes."

He added: "I wouldn't be overly concerned because there's quite often salmonella in raw seafood."

Landauer, which has an annual turnover of more than £26m, told reporters they could save British duty on the consignment by pretending the prawns had been imported from Holland. Seafood imported from America carries 12% duty on the purchase price, but the duty is only 4% if it is brought in direct from India via Holland.

"In respect of the [US] duty, which is considerable, one could invoice at whatever price one wanted," said Brown. "You could do your own invoice." If the consignment was undervalued, less duty would be payable.

A Customs and Excise spokesman said anyone shipping food to Britain from America while claiming it had come straight from India "would be committing fraud".

Landauer offered the mixed consignment, which originally had an American retail value equivalent to £100,000, for £40,000. Irradiation at a plant in The Netherlands would cost £3,500. At British retail prices, the consignment offered a possible profit of more than £100,000.

"I had this offer, like other British wholesalers, and ditched it," said Ken Bell, of Bell International, a major seafood importer. "Such offers are commonplace. But if the product has got filth in it, it has been fished in contaminated waters. Salmonella comes from the bowel of a warmblooded animal, so I could tell you the precise word for it."

Last night Landauer admitted that it regularly offered reject American seafood. "These parcels do come up in the States for re-export," Rinus Verwijs, a Landauer director, said. "We tell our customers it has been rejected. What they do with it then is entirely up to them. We would place them through irradiation for customers."

Additional reporting by Mike Graham, New York.

Appendix C: Who wants irradiation?

C1 Food irradiation and food poisoning
EDM 497, House of Commons Order Paper. April 1989

Mr Frank Cook (Lab.)
Mr Simon Hughes (Lib. Dem.)
Dr Dafydd Elis Thomas (Pl. Cymru)
Mr Churchill (Con.)
Rev Martin Smyth (Unionist)
Mrs Rosie Barnes (SDP)
Mr Alex Salmond (Scot. Nat.)

That this House is of the view that irradiation has no part to play in solving the problem of contamination of foodstuffs arising from food poisoning organisms; notes that the public have perceived the problem correctly as being one of maintaining strict hygiene standards and temperature controls in the production, storage, handling and cooking of foods: notes also a number of well documented cases where irradiation has been used illegally and in violation of world health agencies' advice, deliberately to conceal bacterial contamination on unsaleable foods; believes that the use of irradiation would do further damage to the public confidence in some sections of the food industry; and therefore seeks an assurance that Her Majesty's Government will not remove the current ban on the irradiation of food and will oppose proposals from the European Commission that will require the United Kingdom to permit importation of foodstuffs irradiated for the purpose of reducing bacterial contamination.

Note: sponsered by MPs of all major parties

C2 IOCU resolution on food irradiation

The IOCU General Assembly:

recognising that there are widespread consumer concerns about food irradiation with regard to need, safety, abuse, food quality, nutrition, hygiene, labelling, detection, control of facilities, enforcement and the impact on national and international economies: calls upon IOCU to develop a comprehensive policy paper on food irradiation;

urges IOCU, as a priority, to clarify with the World Health Organisation, the International Atomic Energy Agency and other UN Agencies their role in promoting the acceptance of food irradiation by member governments;

demands a worldwide moratorium on the further use and development of food irradiation until there is a satisfactory resolution of issues of nutrition, safety, labelling and detection;

urges scientists, governments and the food industry, the world over, to research more desirable methods of food preservation; and

realising that food irradiation is permitted in some countries: calls upon IOCU to lobby in the meantime for clear and explicit labelling of irradiated products taking account of practical, and potential problems in date marking, inspection and certification.

ADOPTED BY THE 12TH WORLD CONGRESS OF THE INTERNATIONAL ORGANISATION OF CONSUMERS UNIONS (IOCU), MADRID, SPAIN, 15-20 SEPTEMBER, 1988

C3 Members of the International Consultative Group on food irradiation with dates of joining

Country	Joined
Argentina	1983
Australia	1984
Bangladesh	1983
Canada	1984
Chile	1985
Egypt	1983
Federal Republic of Germany	1984
France	1983
Hungary	1984
India	1984
Indonesia	1986
Iraq	1983
Israel	1983
Italy	1984
Malaysia	1984
Mexico	1984
Netherlands	1983
New Zealand	1985
Pakistan	1986
Philippines	1983
Poland	1985
Syrian Arab Republic	1983
Thailand	1985
Turkey	1983
United States of America	1984
Yugoslavia	1984

Source: New Zealand Ministry of the Environment, 1988

C4 Do Sainsbury's want irradiated food?

Sir John gives cautious welcome to irradiation

SIR John Sainsbury has given his endorsement to food irradiation — but warned there could be problems if it was not handled carefully.

He told a meeting at the Smithfield Show last week that he knew highly qualified people in Europe, the UK and the US who had assessed the facts and believed that it was an important and safe means of food preservation if properly controlled.

"I believe that in time it will serve the interests of the consumer and raise the standards of food preservation in the years ahead," he said.

"But it has got to be handled most carefully and must not be rushed into and that is why we must have the most careful means of reassuring the public."

On the subject of 1992, Sir John revealed Sainsbury's had no plans to open up supermarkets in the rest of Europe.

"Our main business is here. We bought a chain in the US last July and that is a valuable extra dimension to our business."

Before it took that decision the company had considered whether it should be looking at the rest of Europe rather than the US.

But it had concluded that the different nature of each country was such that it would not really be possible to bring trading to those countries as it was in the US where conditions were more closely related to the UK, he said.

Source: *Super Marketing*, 23 December 1988

Appendix D: Contacts and addresses

International contacts

Five regional centres link food irradiation activities around the world by exchanging information and channelling news to groups in their respective regions. Information sent to these five centres will reach all other groups in the network. The centres are:

Europe
London Food Commission, 88 Old Street, London EC1V 9AR

Asia/Pacific
IOCU Asia/Pacific, PO Box 1045, 10830 Penang, Malaysia

Africa
ELCI, PO Box 72461, Nairobi, Kenya

Latin America/Caribbean
AMEDC, Amores 109, Mexico City 03100, Mexico DF

North America
Food and Water Inc, 225 Lafayette Street, New York, NY 10012, USA

Other UK organizations

British Medical Association, BMA House, Tavistock Square, London WC1H 9JP

Consumers' Association, 2 Marylebone Road, London NW1 4DX

Consumers in the European Community, 24 Tufton Street, London SW1

Friends of the Earth, 26/28 Underwood Street, London N1 7JQ

Institute of Environmental Health Officers, Chadwick House, Rushworth Street, London SE1 0QT

National Council of Women, 36 Danbury Street, London N1 8JU

National Federation of Women's Institutes, 39 Eccleston Street, London SW1 9NT

National Union of Townswomen's Guilds, Chamber of Commerce House, 75 Marbourne Road, Birmingham B15

Soil Association, 86/88 Colston Street, Bristol BS1 5BB

Trades Union Congress, 23/25 Great Russell Street, London WC1B 3LS

Vegetarian Society, Parkdale, Dunham road, Altrincham, Cheshire

Food retailers

Retail Consortium, Commonwealth House, 1-19 New Oxford Street, London WC1A 1PA

Argyll PLC, Millington Road, Hayes, Middx UB3 4AY

Asda PLC, Asda House, Britannia Road, Morley, Leeds LS27 0BT

Co-operative Wholesale Society, New Century House, PO Box 53, Corporation Street, Manchester M6 4ES

Dee Corporation, 418 Silbury Boulevard, Milton Keynes MK9 2NB

Mace Line Marketing Ltd, Gerrards House, Station Road, Gerrards Cross, Bucks SL9 8HW

Marks & Spencer PLC, Michael House, Baker Street, London W1A 1DN

Safeway Food Stores Ltd, Beddow Way, Aylesford, Kent ME20 7AT

J. Sainsbury PLC, Stamford House, Stamford Street, London SE1 9LL

Spar, 32/40 Headstone Drive, Wealdstone, Harrow, Middx HA3 5QT

Tesco Stores Ltd, Tesco House, Delamare Road, Cheshunt, Waltham Cross, Herts EN8 9SL

Waitrose Ltd, 4 Old Cavendish Street, London W1A 1EX

Food manufacturers

Food and Drink Federation, 6 Catherine Street, London WC2

Associated British Foods PLC, 68 Knightsbridge, London SW1X 7LR

Booker PLC, Portland House, Stag Place, London SW1E 5AY

Cadbury Schweppes PLC, 1/4 Connaught Place, London W2 2EX

Dalgety UK Ltd, 19 Hanover Square, London W1R 9DA

Hillsdown Holdings PLC, 32 Hampstead High Street, London NW3 1QD

Nestlé, St George's House, Croydon, Surrey CR9 1NR

Northern Foods, Beverley House, St Stephens Square, Hull HU1 3XG

Ranks Hovis McDougall, RHM Centre, Box 178, Alma Road, Windsor, Berks SL4 3ST

Unigate, Unigate House, Western Avenue, London W3 0FH

Unilever PLC, Unilever House, Blackfriars, London EC4P 4BQ

United Biscuits PLC, Grant House, Syon Lane, Isleworth, Middlesex TW7 5NN

Note: A similar list could be drawn up for anywhere in the world

REFERENCES

CHAPTER 2

1 *Food Irradiation: a technique for preserving and improving the safety of food.* World Health Organization, in collaboration with the Food and Agriculture Organization of the United Nations, Geneva, 1988
2 *The Safety and Wholesomeness of Irradiated Foods.* Report of the Advisory Committee on Irradiated and Novel Foods, London, HMSO, 1986
3 *Sources and Effects and Risks of Ionizing Radiation.* Report of the United Nations Scientific Committee on the Effects of Atomic Radiation to the General Assembly, United Nations, New York, 1988
4 J. W. Gofman, *Radiation and Human Health.* Sierra Club Books, San Francisco, 1981
5 R. Bertell, *No Immediate Danger — Prognosis for a Radioactive Earth.* The Women's Press, London, 1985
6 As ref. 1
7 Edward S. Josephson and Martin S. Peterson (eds). *Preservation of Food by Ionizing Radiation.* (3 vols) CRC Press, Florida, Vol. 1 1982, Vols 2 & 3 1983
8 *Wholesomeness of Irradiated Food.* Reports of the Joint Expert Committee of the IAEA/FAO and WHO, World Health Organization, 1977 and 1981
9 Comments from South African and Israeli delegates to the authors at the FAO/IAEA/WHO/ITC-UNCTAD/ITC /GATT International Conference on the Acceptance, Control of and Trade in Irradiated Food, Geneva, 12-16 December 1988
10 Terry Garrett, 'Isotron Profits Expected to Exceed £1 Million Mark'. *Financial Times*, 1 July 1985
11 South Manchester Health Authority, *Appraisal of Catering Methods.* Manchester, 1985
12 As ref. 2
13 As ref. 1
14 As ref. 1

15 *Improvement of Food Quality by Irradiation*. Proceedings of Panel held June 1973. Organized by FAO/IAEA Division of Atomic Energy in Food and Agriculture IAEA/STI/PUB/370, Vienna, 1974
16 P. S. Elias and A. J. Cohen, *Recent Advances in Food Irradiation*. Elsevier Biomedical Press, 1983
17 *Our Daily Bread — Who Makes the Dough*. British Society for Social Responsibility in Science, Agricapital Group, 1978
18 C. Walker and G. Cannon, *The Food Scandal*. Century Publishing, 1985
19 P. S. Elias, *Irradiation of Food*. Environmental Health, Oct. 1982
20 As ref. 7
21 P. S. Elias and A. J. Cohen, *Radiation Chemistry of Major Food Components*. Elsevier Biomedical Press, 1977
22 As ref. 7
23 As ref. 16
24 *Food Irradiation Now*. Symposium in Ede, Netherlands, by Gammaster, 21 Oct 1981. Martinus Nijhof and Dr W. Junk, 1982
25 As ref. 7
26 As ref. 21
27 *The Department of the Army's Food Irradiation Programme — Is it Worth Continuing?* US Government Accounting Office PSAD-78-146, Washington DC: 29 Sept. 1978
28 Department of Health and Human Services, Food and Drug Administration, 21 CFR Part 179, *Irradiation in the Production, Processing and Handling of Food; Final Rule*. 51 Federal Register 13376 at 13385, Washington DC, 18 April 1986
29 Department of Health, *Report of Working Party on Irradiation of Food*. HMSO, London, 1964
30 Food (Control of Irradiation) (Amendment) Regulations 1972 S.I. 1972 No. 205, London, Feb. 1972
31 See answer to Written Question 1398/81 by Mr Narjes on 25 November 1982 to Mr Schmidt and also answer to Mrs Hanna Walz Question 1618/83 given 7 February 1984 in the European Parliament
32 *Wholesomeness of Irradiated Food*. Report of the Joint Expert Committee of the IAEA/FAO and WHO, WHO Technical Report Series No. 451, World Health Organization, 1970
33 *Wholesomeness of Irradiated Food*. Report of the Joint Expert Committee of the IAEA/FAO and WHO, WHO Technical Report Series No. 604, World Health Organization, 1977
34 *Wholesomeness of Irradiated Food*. Report of the Joint Expert Committee of the IAEA/FAO and WHO, WHO Technical Report Series No. 659, World Health Organization, 1981
35 As ref. 34
36 As ref. 1
37 Joint FAO/IAEA Division of Isotope and Radiation Applications of Atomic Energy for Food and Agricultural Development, *Supplement to Food Irradiation Newsletter*. Vol. 12 No. 1, IAEA, Vienna, April 1988

38 L. Pim, *Gamma Irradiation as a Means of Food Preservation in Canada*. Pollution Probe Foundation, Toronto, 1983
39 K. M. Tucker and R. Alvarez, *Comments on Proposed Regulations on Irradiation in the Production, Processing and Handling of Food* (FDA Docket No. 818-0004). Health and Energy Institute, Washington D.C., 1984
40 T. Webb, *Food Irradiation in Britain?* London Food Commission, 1985
41 *UK Government's Advisory Committee on Irradiated and Novel Foods —Terms of Reference and Membership*. DHSS Press Release, 18 May 1982
42 Lord Cockfield's memoranda *Completion of the Internal Market* (COM [85] 310) and *Community Legislation on Foodstuffs* (COM [85] 603). Brussels, 1985
43 *Opinion of the European Parliament on Irradiation of Foodstuffs*. Adopted by European Parliament, Brussels, 10 April 1987
44 International Consultative Group on Food Irradiation, *Marketing Guidelines for Acceptance and Usage of Food Irradiation — Task Force on Marketing and Public Relations of the International Consultative Group on Food Irradiation*. International Atomic Energy Agency, Vienna, September 1986
45 World Health Organization, *Document on Food Irradiation*. Adopted by the FAO/IAEA/WHO/ITC/GATT International Conference on the Acceptance, Control of and Trade in Irradiated Food, Geneva, 16 December 1988
Acceptance Control of and Trade in Irradiated Foods. Conference proceedings, IAEA, Vienna, 1989

CHAPTER 3

1 *Wholesomeness of Irradiated Food*. Report of the Joint Expert Committee of the IAEA/FAO and WHO, World Health Organization, 1981
2 *The Safety and Wholesomeness of Irradiated Foods*. Report of the Advisory Committee on Irradiated and Novel Foods, HMSO, London, 1986
3 *Food Science and Techniques*. Reports of the Scientific Committee for Food, Eighteenth Series, EUR 10840, Commission of the European Communities, 1986
4 *Irradiation of Food*. Translation of a report by a Danish Working Group, Publication 120, Ministry of the Environment, National Food Agency, Søborg, 1986
5 Dept of Health and Human Services, Food and Drug Administration, *Irradiation in the Production, Processing and Handling of Food; Final Rule*. 21 CFR Part 179, Federal Register, 18 April 1986
6 Federal Drug, Food and Cosmetic Act, 21 USC Section 301 *et seq.*, 1986
7 Comments of US delegation to the International Conference on the Acceptance, Control of and Trade in Irradiated Food, Geneva, 12-16 December 1988

8 T. Webb, R. Schilling, B. Jacobson, and P. Babb, *Health at Work?* Health Education Authority, London, 1988
T. Webb, *Changing Perceptions of Risk from Radiation*, MSc Thesis, South Bank Polytechnic, London 1989 (unpublished)
9 *Irradiation of Foodstuffs*. British Medical Association Board of Science, London, March 1987
10 As ref. 2
11 *The Response of the Advisory Committee on Irradiated and Novel Foods (ACINF) to Comments Received on the Report on the Safety and Wholesomeness of Irradiated Foods* (HMSO — ISBN 0 11 321059 0) DHSS, London, 4 February 1988
12 As refs 47 and 55
13 Edward S. Josephson and Martin S. Peterson (eds), *Preservation of Food by Ionizing Radiation*. (3 vols) CRC Press, Florida, Vol. 1 1982, Vols 2 & 3 1983
14 As ref. 2
15 As ref. 11
16 R. S. Hannan, *Scientific and Technical Problems Involved in Using Ionizing Radiations for the Preservation of Food*. Dept of Scientific and Industrial Research, Food Investigation Special Report 61, HMSO, 1955
17 As ref. 2
18 As ref. 2
19 As ref. 13
20 P. S. Elias and A. J. Cohen, *Recent Advances in Food Irradiation*. Elsevier Biomedical Press, 1983
21 P. S. Elias and A. J. Cohen, *Radiation Chemistry of Major Food Components*. Elsevier Biomedical Press, 1977
22 E. Boyland, 'Tumour Initiators, Promoters and Complete Carcinogens', *British Journal of Industrial Medicine*, 47: 716-718, 1984
W. A. Pryor, *Free Radicals in Biology*. New York Academic Press, 1977
23 T. M. Florence, 'Cancer and Ageing — the Free Radical Connection', *Chemistry in Australia*. 50(6): 166-174, 1983
24 As ref. 2
25 As ref. 23
26 As ref. 2
27 As ref. 2
28 As ref. 21
29 As ref. 2
30 J. Barna, 'Compilation of Bioassay Data on the Wholesomeness of Irradiated Food Items', *Acta Alimentaria*. 8 (3): 205-315, Budapest, 1979
31 As ref. 16
32 Statements made to IAEA/FAO/WHO/ITC-UNCTAD/GATT, International Conference on Acceptance Control of and Trade in Irradiated Foods, Geneva, 12-16 December 1988

33 As ref. 13
34 As ref. 13
35 As ref. 30
36 As ref. 30
37 *Outstanding Questions on the Safety of Irradiated Food*. Paper presented to the IAEA/FAO/WHO/ITC-UNCTAD/GATT International Conference on Acceptance Control of and Promotion of Trade in Irradiated Foods by the International Organization of Consumers Unions, Observer group, 1988
See also the Comments on this paper by the ad-hoc group of scientific experts for the WHO at the conference, WHO, Geneva, 1988
38 *Food Irradiation: a technique for preserving and improving the safety of food*. World Health Organization, in collaboration with the Food and Agriculture Organization of the United Nations, Geneva, 1988
39 *The Department of the Army's Food Irradiation Programme — Is it Worth Continuing?* US Government Accounting Office, PSAD-78-146, Washington DC, 29 Sept 1978
40 As ref. 39
41 D. W. Thayer, *Summary of Supporting Documents for Wholesomeness Studies of Pre-cooked (Enzyme Inactivated) Chicken Products in Vacuum Sealed Containers Exposed to Doses of Ionizing Radiation Sufficient to Achieve 'Commercial Sterility'*. US Dept of Agriculture, 19 March 1984
42 Ralston Purina Co., *Irradiation Sterilized Chicken: A Feeding Study in Rats*. Contract No. 53-3K06-1-29, 69, July 1982
43 As ref. 41
44 FDA Bureau of Foods, *Recommendations for Evaluating the Safety of Irradiated Food*. Final Rule, July 1980
45 As ref. 2
46 Frank Cook — letter to Barney Hayhoe, Minister for Health 1986
Michael Meacher — letter to Norman Fowler, Secretary of State for Health and Social Services 1986
Brynmor John — letter to Michael Jopling, Minister for Agriculture, Fisheries and Food 1986
47 Report of the Working Party on Food Irradiation to the Ministry of Health, HMSO, London, 1964
48 *Wholesomeness of Irradiated Food*, Reports of the Joint Expert Committee of the IAEA/FAO and WHO. World Health Organization, 1977 and 1981
49 As ref. 3
50 As ref. 48
51 Survey of reports cited by the IAEA/FAO/WHO — JECFI Committees by the Food Chain Coalition, Toronto 1987
52 Conversation with Dieter Ehlermann at the International Conference on the Acceptance Control of and Trade in Irradiated Food, Geneva, December 1988

53 Conversation between Tony Webb, Peter Elias and Dr Vijayalaxmi at Consensus Conference on Food Irradiation in Copenhagen, 22-24 May 1989
54 *Use of Ionizing Radiation*. Report of the House of Representatives Standing Committee on Environment, Recreation and the Arts, Canberra, November 1988
55 World Health Organization, *Wholesomeness of Irradiated Food with special reference to wheat, potatoes and onions*. Report of a Joint FAO/IAEA/WHO Expert Committee, WHO Technical Report Series No. 451, Geneva, 1970
56 Vijayalaxmi and K. Visweswara Rao, 'Dominant Lethal Mutations in Rats Fed on Irradiated Wheat, *Int. Journal of Radiation Biology*, 29: 93-98, 1976
57 Vijayalaxmi, 'Genetic Effects of Feeding Irradiated Wheat to Mice'. *Canadian Journal of Genetic Cytology* 18: 231-238, 1976
58 As refs 56 and 57
59 As ref. 57
60 Vijayalaxmi, 'Immune Response in Rats Fed Irradiated Wheat'. *British Journal of Nutrition* 40: 535-541, 1978
61 Vijayalaxmi and G. Sadasivan, 'Chromosomal Aberrations in Rats Fed Irradiated Wheat'. *International Journal of Radiation Biology* 27: 283-285, 1975
62 As ref. 57
63 Vijayalaxmi, 'Cytogenetic Studies in Monkeys Fed Irradiated Wheat'. *Toxicology* 9: 181-184, 1978
64 C. Bhaskharam and G. Sadasivan, 'Effects of Feeding Irradiated Wheat to Malnourished Children'. *American Journal of Clinical Nutrition* 28: 130-135, 1975
65 K. P. George, R. C. Chaubey, K. Sundaram and Gopal-Ayengar, 'Frequency of Polyploid Cells in the Bone Marrow of Rats Fed Irradiated Wheat'. *Fd Cosmet. Toxicology* 14: 289-291, 1976
66 O. S. Reddi, P. P. Reddy, D. N. Ebenezer and N. V. Naidu, 'Lack of genetic and cytogenetic effects in mice fed irradiated wheat' *International Journal of Radiation Biology* 31: 589-601, 1977
67 Tesh, Davidson, Walker, Palmer, Cozens, and Richardson, *Studies in Rats Fed a Diet Incorporating Irradiated Wheat*. International Project in the Field of Food Irradiation, IFIP-R45, Karlsruhe, 1977
68 D. Anderson, M. J. L. Clapp, M. C. E. Hodge and T. M. Weight, 'Irradiated Laboratory Animal Diets. Dominant Lethal Studies in the Mouse'. *Mutation Research* 80: 333-345, 1981
69 H. W. Renner, 'Chromosome Studies in the Bone Marrow of Chinese Hamsters Fed a Radiosterilized Diet'. *Toxicology* 8: 213-222, 1877
70 Jacobs et al., *Nature* 193: 591, 1962
71 E. Boyland, 'Tumour Initiators, Promoters and Complete Carcinogens, *British Journal of Industrial Medicine* 47: 716-718 1984
72 Reply to the criticisms of the Report of the Indian government by Drs Kesavan and Sukhatme submitted by NIN 1977

73 Statements by Martin Welt, Radiation Technology Inc, to conference on Alternative Uses for Nuclear Energy: Focus on Food Irradiation. Joint American Nuclear Society/European Nuclear Society International Conference, 13 November 1984
74 Vijayalaxmi and S. G. Srikantia, 'A Review of the Studies on the Wholesomeness of Irradiated Wheat, Conducted at NIN, India', in press for publication in *Radiation Physics and Chemistry*, 1989
75 P. C. Kesavan, 'Indirect effects of radiation in relation to food preservation: facts and fallacies'. *Journal of Nuclear Agricultural Biology* 7: 93-97, 1978
76 As ref. 74
77 Report of Drs Kesavan and Sukhatme to the Indian Government 1977. Unpublished but was widely circulated by the IAEA at the International Conference on Acceptance Control of and Trade in Irradiated Food, Geneva 1988
78 As ref. 74
79 A. Brynjolfsson, 'Wholesomeness of Irradiated Foods: a review'. *Journal of Food Safety* 7: 107-126, 1985
80 Claim by Peter Elias Director of the IFIP
81 As ref. 77
82 O. Frota-Pessoa, N. R. Ferreira, M. B. Pedroso, A. M. Moro, P. A. Otto, D. A. F. Chamone and L. C. DaSilva, 'A study of chromosomes of lymphocytes from patients treated with hucanthone'. *J. Toxicol. Env. Health* 1: 305-307, 1975
83 As ref. 74
84 As ref. 75
85 See also D. McPhee and W. Hall, 'An analysis of the safety of food irradiation: Genetic Effects' in *Use of Ionizing Radiation*. Report of the House of Representatives Standing Committee on Environment, Recreation and the Arts, Canberra, November 1988
86 As ref. 2
87 As ref. 9
88 51 Fed, Reg. 13376 at 13385, 18 April 1986
'FDA Erred Citing FAO/WHO Review of Indian Irradiated Wheat Study'. *Food Chemical News* 28: 17, November 1986
89 As ref. 67
90 Vijayalaxmi, *A Critical Evaluation of the Studies Conducted on the Wholesomeness of Freshly Irradiated Wheat*. In press 1989
91 As ref. 90
92 K. Tucker and M. Rabinowitz, *More False Promises from the Nuclear Industry*. Health and Energy Institute, Washington DC, December 1985
93 Dai Yin, *An Introduction of Safety Evaluation on Irradiated Foods in China*. Paper to FAO/IAEA Seminar for Asia and the Pacific on the Practical Application of Food Irradiation, Shanghai, China, 7-11 April 1986
94 Li Hao et al., 'Feeding Trial of Gamma-Ray Irradiated Potato in Human Volunteers'. *Food Hygiene Research* 2(2): 32-37, 1984

95 Hou Yn Hua et al., 'Feeding Trial of Gamma-Ray Irradiated Rice in Human Volunteers'. *Food Hygiene Research* 2(2): 18-24, 1984
96 He Zhiqian et al., 'Study on Safety of Irradiated Mushroom to Human Body'. *Food Hygiene Research* 2(2): 43-49, 1984
97 Zhang Yan et al, 'Feeding Trial of Gamma-Ray Irradiated Meat Products in Human Volunteers'. *Food Hygiene Research* 2(2): 69-74, 1984
98 Li Juesen et al., 'Feeding Trial of Gamma-Ray Irradiated Peanut in Human Volunteers'. *Food Hygiene Research* 2(2): 56-61,1984
99 As ref. 93
100 As ref. 38
101 'Safety of 35 Kinds of Irradiated Human Foods'. *Chinese Medical Journal* 100: 715-718, 1987
102 Consensus Conference on Food Irradiation held 22-24 May in the Landsting Hall, Christiansborg, organized by the Danish Board of Technology (Teknologienaevnet) in co-operation with the Research Committee of the Danish Parliament, Teknologienaevnet, Copenhagen, 1989
103 *Ionizing Energy in Food Processing and Pest Control: 1. Wholesomeness of Food Treated With Ionizing Energy*, Council for Agricultural Science and Technology, Report No. 109, USA, July 1986
104 As ref. 2
105 As ref. 48
106 As ref. 37
107 As ref. 1
108 Particularly Electron Spin Resonance (ESR) and Chemi/Fluoro-Luminescence techniques — see Chapter 4
109 As ref. 11
110 As ref. 38
111 Evidence of Fiona Cummings to the Australian House of Representatives Standing Committee on Environment, Recreation and the Arts, 1987
112 Analysis by Leah Bloomfield, presented to IOCU Regional Consultation on Food Irradiation, Canberra, October 1988, proceedings published by NCSFI/AFCO/CC, Canberra & Melbourne, 1988
113 As ref. 11
114 As ref. 37
115 As ref. 11
116 As refs. 37 and 38
117 As ref. 30
118 As ref. 48
119 As ref. 20

CHAPTER 4

1 P. S. Elias and A. J. Cohen, *Radiation Chemistry of Major Food Components*. Elsevier Biomedical Press, 1977

2 Edward S. Josephson and Martin S.Peterson (eds), *Preservation of Food by Ionizing Radiation*. (3 vols) CRC Press, Florida, Vol. 1 1982, Vols 2 & 3 1983
3 M. D. Rankin in *Food Industries Manual*, London
4 As ref. 2
5 As ref. 2
6 National Advisory Committee on Nutrition Education, Health Education Council, London, 1983
Committee on the Medical Aspects of Food Policy (COMA), *Diet and Cardiovascular Disease*, HMSO, London, 1984
Nutrition and Your Health — Dietary Guidelines for Americans. US Department of Agriculture and US Department of Health and Human Services, 1985
7 As ref. 2
8 As ref. 2
9 P. S. Elias and A. J. Cohen, *Radiation Chemistry of Major Food Components*. Elsevier Biomedical Press, Amsterdam and New York, 1977.
10 P. S. Elias and A. J. Cohen, *Recent Advances in Food Irradiation*. Elsevier Biomedical Press, Amsterdam and New York, 1983.
11 W. Heiman, *Fundamentals of Food Chemistry*. Ellis Horwood, Chichester, UK, 1980
12 R. W. Wenlock et al. *The Diets of British Schoolchildren*. Department of Health and Social Security, HMSO, London, 1986
13 Hearing of the US Senate Select Committee on Hunger, Washington DC, 4 April 1986
M. Jacobson, *The Complete Eater's Digest and Nutritional Scoreboard*. Anchor/Doubleday, Garden City NY, 1986
14 *The National Dietary Survey of Adults: 1983 — Nutrient Intakes*. Federal Department of Community Services and Health, and the National Heart Foundation, Canberra, April 1988
15 Press Release on 103rd Session of the National Health and Medical Research Council, Hobart, June 1987
16 R. W. Smithells et al. 'Further Experience of Vitamin, Supplementation for Prevention of Neural Tube Defects Recurrencies' *Lancet*, 1027-1031, 1983
17 M. H. Soltan and D. M. Jenkins, 'Maternal and Foetal Plasma Zinc Concentration and Foetal Abnormality.' *Brit. J. Obstet. Gynaecol*, 89: 56-58, 1982
18 *The Safety and Wholesomeness of Irradiated Foods*. Report of the Advisory Committee on Irradiated and Novel Foods, HMSO, London, 1986
19 As ref. 11
20 As ref. 2
21 N. Raica, J. Scott, and W. Neilson 'The Nutrional Quality of Irradiated Foods', *Radiation Research Reviews*, 3: 447, 1972
22 Statement of Sir Arnold Burgen, DHSS Press Conference, 10 April 1986

23 As ref. 18
24 Titlebaum, Dubin and Doyle, 'Will Consumers Accept Irradiated Foods?' *Journal of Food Safety* 5: 219-228, 1983
Bruhn, Shultz and Sommer, 'Attitude Change Toward Food Irradiation Among Conventional and Alternative Consumers.' *Food Technology*, Jan. 1986
Survey conducted by the UK Consumers' Association, Nov. 1986
Food Irradiation — Omnibus Research by Marplan Ltd, conducted for the London Food Commission, Jan. 1987
Brand Group, *Irradiated Seafood Products: A Position Paper for the Seafood Industry*. Prepared for the National Marine Fisheries Service Washington DC, Jan. 1986
25 Statement to IAEA Research Project on Food Irradiation for the Asia Pacific Region held Bangkok, October 1988
Statement to the FAO/IAEA/WHO/ITC-UNCTAD/GATT International Conference on the Acceptance Control of and Trade in Irradiated Food, Geneva, 12-16 December 1988
26 *Food Irradiation: a technique for preserving and improving the safety of food*. World Health Organization, in collaboration with the Food and Agriculture Organization of the United Nations, Geneva, 1988
27 Consensus Conference on Food Irradiation Final Document. Conclusions of the Question Panel at the Consensus Conference on Food Irradiation held 22-24 May 1989 in the Landsting Hall at Christiansborg. Organized by the Danish Board of Technology (Teknologienaevnet) in co-operation with the Research Committee of the Danish Parliament, Teknologienaevnet, Copenhagen, 1989
28 W. P. T. James in collaboration with A. Ferro-Luzzi, B. Isakssow & W. B. Szostak, *Healthy Nutrition: Preventing nutrition-related diseases in Europe*. WHO Regional Publications, European Series, No. 24, 1988
29 *Food Irradiation*. A report to the New Zealand Ministry of the Enviroment, Wellington, 1987

CHAPTER 5

1 Stephen Leather, 'We Broke the Law with Gamma Ray Prawns, Say Food Firm.' *Daily Mail*, London, 3 March 1986
2 *4 What It's Worth*, London, Thames TV to Channel 4, 8 April 1986
3 *The Safety and Wholesomeness of Irradiated Foods*, Report of the Advisory Committee on Irradiated and Novel Foods, HMSO, London, 1986
4 Early Day Motion No. 714, *Irradiated Food: Illegal Importation*. House of Commons Order Paper, 9 April 1986
5 As ref. 2
6 Per Axel Janson, *Dagens Nyheter*. Stockholm, 8 August 1986
7 Letter from Peggy Fenner, Parliamentary Secretary to the Ministry of Agriculture Fisheries and Food, to the Director of the London Food Commission, 1986

8 As ref. 2
9 J. B. Nielsen, 'Danish firm is sending tainted mussels to Holland to be irradiated'. *Information*, Copenhagen, 3 February 1987
O. Lindboe, Reporting of irradiated mussels. *Politiken*, Copenhagen, 3 February 1987
'NOAH erstattet Anzeige gegen Roemoe Muselinge Kompagniet'. *Der Nordschleswiger*, Aabenraa, 3 February 1987
J. B. Nielsen. 'Will not destroy irradiated mussels'. *Information* (Copenhagen), 20 February 1987; 'Fines are to be expected for irradiated mussels'. *Aktuelt* Copenhagen, 25 February 1987; 'Irradiated mussels on Roemoe are destructed'. *Vestksten* Esbjerg, 10 April 1987; 'Beschluss. Bestrahlte Muscheln vernichten oder verfuttern'. *Der Nordschleswiger,* Aabenraa, 10 April 1987
J. Toft, 'Irradiated common mussels', *NOAH-Bladet* 107: 9-10, Copenhagen, April 1987
10 Comments by Gammaster to Liz Leigh, Journalist, *Sunday Times*, during visit to Gammaster 1986
Comments by Jan Leemhorst, Gammaster, to Food Irradiation in Europe seminar organized by the London Food Commission, London, April 1988
11 *Food Irradiation, Abuses and Consumer Concerns*. Paper presented by the International Organization of Consumers Unions to the International Conference on the Acceptance Control of and Trade in Irradiated Foods, Geneva, 13 December 1988
12 *Japanese Consumers Want No Food Irradiation*, Food Irradiation Network, Japan, and Consumers Union of Japan, Sagamihara-city, 1989
13 J. Savagian and D. Mosgofian, *Basic information regarding Quaker Oats Company's use of irradiated mushrooms*. New York Public Interest Research Group, Inc. New York, and National Coalition to Stop Food Irradiation, San Francisco, October 1988
14 Early Day Motion No. 950, *Illegal Food Irradiation*. Hansard, London, 9 June 1989
15 Personal communication: Notes on conversation between journalist Peter Spinks and Irene DeBruijne, 9 May 1989
16 Barrie Penrose, 'Illegal Trade in spoiled seafood uncovered'. *Sunday Times*, London, 6 August 1989
17 Statement by Caroline Jackson MEP in the European Parliament debate, March 1987.
18 As ref. 16
19 Information supplied by Dr Engel, USDA, Washington DC, December 1986
20 Office of Population Census and Surveys, OPCS Monitor, London, 1988
21 T. Webb, *Food Irradiation and Food Poisoning*, A briefing paper by the London Food Commission, London, March 1989
22 *The Safety and Wholesomeness of Irradiated Foods*. Report of the Advisory Committee on Irradiated and Novel Foods, HMSO, London, 1986

23 *Factors Influencing the Economical Application of Food Irradiation.* Proceedings of a Panel held Venice, June 1971, organized by FAO/IAEA Division of Atomic Energy on Food and Agriculture, IAEA STI/PUB/331, 1973
24 Edward S. Josephson and Martin S. Peterson (eds), *Preservation of Food by Ionizing Radiation.* (3 vols) CRC Press, Florida, Vol. 1 1982, Vols 2 & 3 1983
25 As ref. 24
26 M. W. Eklund, 'Significance of Clostridium botulinum in fishery products preserved short of irradiation'. *Food Technology* 36 (12): 107, 1982
27 H. Julius, *Food Irradiation Fact Sheet: The Bacterial Risks of Irradiated Food.* Unpublished; available from the Citizens Concerned About Food Irradiation, PO Box 236 Red Hill, Queensland 4059, Australia
28 *Scheme and Critical Variables for a Limited Study on the Effects of Vacuum Packaging and Irradiation on the Outgrowth and Toxin Production of Clostridium Botulinum in Pork Loins.* US Department of Agriculture, Food Safety Inspection Service, Washington DC, June 1986
29 R. S. Hannan, *Scientific and Technical Problems Involved in Using Ionizing Radiations For the Preservation of Food* Dept of Scientific and Industrial Research, Food Investigation Special Report 61, HMSO, 1955
30 P. S. Elias and A. J. Cohen, *Radiation Chemistry of Major Food Components.* Elsevier Biomedical Press, Amsterdam and New York, 1977
31 As ref. 24
32 As ref. 3
33 I. R. Grant and M. F. Patterson, 'A novel radiation-resistant Deinobacter sp. isolated from irradiated pork.' Letters in *Applied Microbiology* 8: 21-24 1989
34 Codex Alimentarius Commission, *The Microbiological Safety of Irradiated Food.* Report of a meeting of the Board of International Committee on Food Microbiology and Hygiene of the International Union of Microbiological Societies with participation of WHO, FAO and IAEA in Copenhagen, 16 December 1982. FAO/WHO CX/FH 83/9, Rome 1983.
35 S. A. Rose, N. K. Modi, H. S. Tranter, N. E. Bailey, M. F. Stringer & P. Hambleton, 'Studies on the irradiation of toxins of Clostridium botulinum and Staphylococcus aureus'. *Journal of Applied Bacteriology*, 65: 223-229, 1988
36 A. C. Pier, M. E. McLoughlin, 'Mycotoxic suppression of immunity'. In *Tricothecenes and other mycotoxins*, J. Lacey (ed), 507-519, John Wiley, Chichester, 1985
37 Bullerman et al., 'Use of Gamma Irradiation to Prevent Aflatoxin Production in Bread'. *Journal of Food Science* 1238, 1973
E. Pryadorshini, P. B. Tupule, 'Effects of Graded Doses of Irradiation

on Aflatoxin production by Aspergillus parasiticus in Wheat'. *Cosmet. Toxicology* 505, 1979
'Aflatoxin Production on Irradiated Foods'. *Cosmet. Toxicology* 293, 1976
38 A. F. Schindler, A. N. Abadie and R. E. Simpson, 'Enhanced Aflatoxin Production by Aspergillus flavus and Aspergillus parasiticus after Gamma Irradiation of the Spore Inoculum'. *Journal of Food Protection* 43 (1): 7-9, January 1980
39 As ref. 37
40 H. Julius, *Food Irradiation and Moulds: A Time Bomb.* Submission to the House of Representatives Committee on the Environment, Recreation and the Arts, Enquiry into Irradiation of Food in Australia, Canberra, 1987
41 See recommendations in *Food Irradiation — An Inquiry by the Australian Consumers' Association*. Undertaken at the request of the Commonwealth Minister of Health, A.C.A., Sydney, April 1987
42 As ref. 41
43 Discussions with Craig Samms, Whole Earth Foods, UK, 1986
44 For fuller treatment see London Food Commission, *Food Adulteration: and how to beat it.* Unwin Hyman, London, 1988
45 T. Roberts, Statement to ICGFI meeting on Microbiological Standards for Irradiated Food, Geneva, June 1989
A. Holmes, Leatherhead Food Industries Research Association, statements on various radio programmes where compulsory irradiation of all chicken has been advocated.
46 Contribution from the floor during one-day seminar on Food Irradiation organized by Institute of Environmental Health Officers, London, 1987
47 Dr K. Bogel, Veterinary and Public Health Unit of the World Health Organization, to ICGFI meeting on Microbiological Standards for Irradiated Foods, Geneva, June 1989
48 Statement of poultry producer to press briefing on food irradiation organized by the Royal Society, London, Sept. 1987
49 Statement by representative of D. B. Marshall to meeting on food irradiation organized by Lothian Regional Council, Linlithgow, Scotland, 1987
50 Statement by Mrs Edwina Currie, December 1988
51 L. Curtin, *Economic Study of Salmonella Poisoning and Control Measures in Canada*. Food Markets Analysis Division, Marketing and Economics Branch, Agriculture Canada, August 1984
52 *Salmonellosis Control: The Role of Animal and Product Hygiene*. World Health Organization, Geneva, December 1988
53 B. F. Yule, J. C. M. Sharp, G. I. Forbes and A. F. MacLeod. '*Prevention of poultry-borne salmonellosis by irradiation: costs and benefits in Scotland*'. Bulletin of World Health Organization 66 (6): 753-758, WHO, Geneva, 1988
54 International Consultative Group on Food Irradiation, Task force meeting on the use of irradiation to ensure hygenic quality of food, 14-18 July 1986. World Health Organization, Geneva, 1987

55 J. Erlichman, 'UK's foul abattoirs anger Europe'. *The Guardian*, London, 19 March, 1989
56 World Health Organization, *Document on Food Irradiation*. Adopted by the FAO/IAEA/WHO/ITC-UNCTAD/GATT International Conference on the Acceptance Control of and Trade in Irradiated Food, Geneva, 16 December 1988
57 Comments by R. Dawson, Senior Officer, Food Quality and Consumer Protection Group, FAO, Rome to meeting on Microbiological Standards for Irradiated Food, Geneva, May 1989
58 Statements by Ken Bell MBE, Managing Director, Ken Bell International, on the consumer programme *4 What it's Worth*, April 1986; at launch of the LFC Food Irradiation Campaign, February 1987; and in numerous press, radio and TV interviews
59 As ref. 44
60 As ref. 24
61 *The Response of the Advisory Committee on Irradiated and Novel Foods (ACINF) to Comments Received on the Report on the Safety and Wholesomeness of Irradiated Foods* (HMSO — ISBN 0 11 321059 0) Department of Health and Social Security, London, 4 February 1988
62 Data reported to meeting of the Asia/Pacific Region Research Project on Food Irradiation (RPFI phase III), organized by the International Atomic Energy Agency in Bangkok, Thailand, October 1988
63 Melvin Couey, USDA Research Branch, Hilo, reported in *Hawaii Tribune Herald* 2 April 1987
64 IAEA News Features, No. 5, IAEA, Vienna, December 1985
65 'Irradiated Strawberries'. *The Food Magazine*, Vol. 1, No. 5, (Summer 1989.)
66 As ref. 62
67 Dr F. E. Peters, former head of Australian Goverment Analytical Laboratories, personal communication, 1989
68 Rosanna Mentzer Morrison and Tanya Roberts, *Food Irradiation: New Perspectives on a Controversial Technology — A Review of Technical, Public Health and Economic Considerations*. Report prepared for the Office of Technology Assessment, Congress of the United States, by the Economic Research Service of the US Dept of Agriculture, December 1985
69 U. Bloch-Von Blottnitz, Report drawn up on behalf of the Committee on the Environment, Public Health and Consumer Protection on the proposal from the European Commission for a Council Directive on the approximation of the laws of the Member States concerning foods and food ingredients treated with ionizing radiation (COM (88) 654-Doc.C 2-280/88), European Parliament Document A 2-69/89. 5 April 1989
70 See comments of R. Dawson, in report by R. Stevens, Public Analyst, London Borough of Southwark and IOCU representative to ICGFI meeting on Microbiological Standards for Irradiated Foods, Geneva, June 1989

71 M. Beyers, 'The retardation by gamma irradiation of greening in potatoes (Solanum tuberosum) exposed to fluorescent lighting'. *Agroplantae* 13: 29-38, 1982
72 S. C. Morris and T. H. Lee 'Toxicity and Teratogenicity of Solanceae glycoalcaloids particularly those of the potato (Solanum tuberosum); a review'. *Food technology Aust*. 36: 118-24, 1984
73 S. Schwimmer and W. J. Weston 'Chlorophyll formation in potato tubers as influenced by gamma irradiation and by chemicals'. *Am. Potato J*. 35: 534-42, 1958
74 M. T. Wu and D. K. Salunkhe 'Effect of gamma irradiation on wound-induced glycoalkaloid formation in potato tubers.' *Lebensm. Wiss. Technol*. 10: 141-4, 1977
75 L. Pim. *Gamma Irradiation as a Means of Food Preservation in Canada*. Pollution Probe Foundation, Toronto, 1983
76 As Ref. 12
77 Alan Holmes, Director of Leatherhead Food Industries Research Association, personal communication
78 As ref. 62
79 As ref. 12
80 Pete Snell and Kirsty Nicol, *Pesticide Residues and Food — a case of real control*. London Food Commission, 1986
81 Melanie Miller, *Danger! Additives at Work*. London Food Commission, 1985
82 As ref. 24
83 Statements to Tony Webb and Beverley Sutherland-Smith at meeting with Coles-Meyer, Melbourne, May 1988
84 As ref. 24
85 As ref. 81
86 N. H. Proctor and J. P. Hughes, *Chemical Hazards in the Workplace*. Lippincott, Philadelphia, 1978
87 As ref. 24
88 As ref. 24
89 M. Windholz, ed, *The Merck Index — An Encylopedia of Chemicals and Drugs* (9th Edition). Merck and Co, Rahway, New Jersey, 1976
90 As ref. 44
91 Judith A. DeCava, 'Facts About Food Irradiation'. *Journal of the National Academy of Research Biochemists*, July 1986
92 As ref. 68
93 *Food Irradiation: a technique for preserving and improving the safety of food*. World Health Organization, in collaboration with the Food and Agriculture Organization of the United Nations, Geneva, 1988
94 *Radiation and Health*. Proceedings of a Conference organized by The National Radiological Protection Board and Friends of the Earth, Hammersmith Hospital, London, in September 1987, J. Wiley, London, 1987
95 National Radiological Protection Board, *Interim Guidance on the Implications of Recent Revisions of Risk Estimates and the 1987 Como Statement*. NRPB-GS9, HMSO, London, November 1987

96 T. Webb, *Changing Perceptions of Risk from Radiation — The case for reducing the risk and some implications for the energy sector.* MSc in Energy Resources Management, South Bank Polytechnic, London, January 1989 (unpublished)
97 *Annals of the International Commission on Radiological Protection — ICRP Report 26*. Pergamon Press, Oxford, 1977
98 Brief of the Ontario Hydro Exployees Union to the Atomic Energy Control Board on Proposed changes to Regulations under the Atomic Energy Control Act. OHEU Toronto, Canada, Jan. 1984
99 51 Federal Register 23612, Washington, 30 June 1986
100 US NRC v. Radiation Technology 519 Fed. Supp. 1266. (D.NJ 1981)
101 *Food Irradiation Alert*. Newsletter of the National Coalition to Stop Food Irradiation, Issue 12, Vol. 3, No. 1, San Francisco, December 1988
102 Dave Harada-Stone, 'N. J. company looks into isle irradiation plant to sterilize imported medical gear'. *Hawaii Tribune-Herald*, 11 July 1989
See also publicity material distributed by Dr Andrew Welt, Alpha Omega Technology, Hawaii, June 30 1989
103 US AEC order to Isomedix Ind. re: Byproduct Material Licences Nos 29-15364-01 and 29-15364-01 14 June 1974
US NRC Notice of violation and letter to Isomedix 4 September 1979
NRC Inspection Report for Isomedix, Docket No. 30-08985, 7 July 1981
Letter from Thomas T. Martin, US NRC to Isomedix, Docket No. 30-19752, 18 October 1984
104 US NRC Order Modifying the Licence of International Nutronics, Licence No. 29-13848-01, 1 Nov. 1983
NRC print-out of accidents, item no. 82-0216, reported 25 October 1982
US NRC Order Modifying the Licence Effective Immediately of International Nutronics, Licence No. 29-13848-0, 30 Jan. 1984
US NRC Response to Questions Concerning Immediately Effective Order Issued to International Nutronics Inc, 14 June 1984
104aLetter from A. M. Dollar, Supervisor, Hawaii Development Irradiator, to H. Brook, AEC, 5 April 1967. Letter from R. M. Baltzo to Y. Kitagawa, Hawaii Department of Agriculture, 9 June 1980 'Radiological Contamination and Hazard at the Hawaiian Developmental Irradiator, 1968 through 1980'. Food Irradiation Response Factsheet, 1986
105 J. L. Setser, *Summary — First Interim Report of the RSI Incident Evaluation Task Force*. Georgia Department of Natural Resources, Atlanta, June 1989
106 Gamma Radiation Accidents reported by the IAEA. Source: Jamaica Irradiation Technology Feasibility Study for the Canadian International Development Agency (CIDA), Ottawa, 1988

107 As ref. 106
108 As ref. 106
109 As ref. 106
110 As ref. 106
111 Olivia Ward, 'Canadian-made equipment cited in El Savador irradiation mishap. Three injured in irradiation accident.' *Toronto Star*, 9 July 1989
112 House of Representatives Standing Committee on Environment, Recreation and the Arts, Official Hansard Report of public hearing, Sydney, 4 March 1987
113 'The Irradiation Industry Hall of Shame'. In *Food Irradiation Alert*, The Newsletter of the National Coalition to Stop Food Irradiation, Issue 14 Vol. 3 No. 3, San Francisco, June 1989
114 As ref. 93
115 European Committee for Food of the International Union of Foodworkers ECF/IUF statement on Food Irradiation, Geneva, 1988 Asia Pacific Region resolution on Food Irradiation, adopted Seoul, 10-12 May 1989

CHAPTER 6

1 *Draft Marketing Guidelines for Acceptance and Usage of Food Irradiation*. Print-out from computer disk, Recommended Marketing Plan from the task force on Marketing/Public Relations established by the International Consultative Group on Food Irradiation based on meeting held 19 September 1986
2 International Consultative Group on Food Irradiation, *Marketing Guidelines for Acceptance and Usage of Food Irradiation*. Task Force on Marketing and Public Relations of the International Consultative Group on Food Irradiation 15-19 September 1986, International Atomic Energy Agency, Vienna, 1987
3 As ref. 1
4 As ref. 1
5 As ref. 2
6 *In Point of Fact* No. 40, World Health Organization, Geneva, 1987
7 *Food Irradiation: a technqiue for preserving and improving the safety of food*. World Health Organization in collaboration with the Food and Agriculture Organization of the United Nations, Geneva, 1988
8 Meeting of the Asia Pacific Region Research Project on Food Irradiation (RPFI Phase III), held at the Office for Atomic Energy for Peace (OAEP), Bangkok, Thailand, October 1988
9 T. Webb, Report to IOCU on the RPFI Phase III meeting held Bangkok, Oct, 1988, IOCU, Penang, 1988
10 Letter from Tony Webb, IOCU representative to RPFI Phase III meeting Bangkok, Thailand, to Paisan Loaharanu, IAEA, November 1988
11 Action alert issued by the International Organization of Consumers Unions, Penang, October 1988

12 The initiative was reported to RPFI Phase III meeting Bangkok, October 1988 by the delegate from Pakistan. The quote was from Paisan Loaharanu of the IAEA who was secretary to the meeting.
13 Statements to RPFI Phase III meeting, Bangkok, Oct. 1988 and to the International Conference on Acceptance Control of and Trade in Irradiated Foods, Geneva, December 1988
13a Ann Usher, 'Ray of hope or leap in the dark?' *The Nation*, Bangkok, 17 August 1989
14 A full list of these companies is provided in T. Webb, T. Lang and K. Tucker, *Food Irradiation — Who Wants It*? Thorsons, USA, 1987
15 For complete analysis of this media hype and the facts about industry and consumer opinion at the time see T. Webb and T. Lang, *Food Irradiation: the Facts*, Thorsons, Wellingborough, 1987
16 *A Revolution in Food Preservation — A consumer guide to food irradiation*. Food and Drink Federation, London, 1986
17 James Erlichman, *The Guardian*; Jane Dawson, *British Medical Journal*, September 1987
18 American Medical Association Policy Statement, Washington DC, 1986
19 Letter from W. T. McGivney, Director of Technology Assessement, AMA, to J. V. Kelly, New Jersey Assemblyman, 1988
20 *Irradiation of Foodstuffs*, British Medical Association Board of Science, London, March 1987
21 World Health Organization, *Document on Food Irradiation*. Adopted by the FAO/IAEA/WHO/ITC-UNCTAD/GATT International Conference on the Acceptance, Control of and Trade in Irradiated Food, Geneva, 16 December 1988
22 *Use of Ionizing Radiation*. Report of the House of Representatives Standing Committee on Environment, Recreation and the Arts, Canberra, November 1988
23 M. Collins, Chairperson, Report of a Parliamentary Committee on Food Irradiation, Ottawa, 1987
24 *Food Irradiation*. A report to the New Zealand Ministry of the Environment, Wellington, 1987
25 *Consensus Conference on Food Irradiation Final Document*, Conclusions of the Question Panel at the Consensus Conference on Food Irradiation held 22-24 May 1989 in the Landsting Hall at Christiansborg organized by the Danish Board of Technology (Teknologienaevnet) in co-operation with the Research Committee of the Danish Parliament, Teknologienaevnet, Copenhagen, 1989
25a J. Taylor, *Consumer Views on the Acceptance of Irradiated Foods*. Keynote address to International Conference on the Acceptance Control of and Trade in Irradiated Food, Geneva, 12-16 Dec. 1988
25b *Acceptance Control of and Trade in Irradiated Foods*. Conference proceedings, IAEA, Vienna, 1989
25c As ref. 7
26 Comments on IOCU's paper 'Outstanding Questions on the Safety of Irradiated Foods'. Ad hoc scientific group, International Conference

on the Acceptance of Irradiated Food, Geneva, 1988
Outstanding Questions on the Safety of Irradiated Food, IOCU, Penang, Malaysia, 1988

27 L. Bloomfield, 'Priority issues on safety and abuse of irradiation.' Oral presentation to the International Conference on the Acceptance Control of and Trade in Irradiated Food, Geneva, 13 December 1988

28 Report by R. Stevens, Public Analyst, London Borough of Southwark and IOCU representative to ICFGI meeting on Microbiological Standards for Irradiated Foods, Geneva, June 1989

29 ECF Position on Document Com (85) 603 final of the EC dated November 1985, Completion of the Internal Market: Community Legislation on foodstuffs; European Committee of Food Catering and Allied Workers Unions within the International Union of Foodworkers (ECF-IUF), 12 June 1986

30 Statement to London Food Commission by Debbie Berkowitz, Health and Safety Director, Food and Allied Service Trades, Dept of AFL-CIO, March 1987

31 Joint Press Statement of the Consumers in the European Community Group, The Institute of Environmental Health Officers and the Retail Consortium, May 1989

32 For fuller discussion see T. Webb and T. Lang, *Food Irradiation — The Facts*. Thorsons, Wellingborough, 1987

33 As ref. 16

34 Statement of FDF President R. Buckland to James Erlichman, *The Guardian*, London, 20 February 1987

35 As ref. 32

36 e.g. British Mushroom Growers Association, 1987

37 NFU statement to European Food Irradiation Network Organizing Meeting, Brussels, 1988

38 See ref. 31

39 Evidence of the Spice Trade Federation to the House of Commons Select Committee on Food Safety, 1989

40 For examples see letters from Donna Elliott, Manager — Customer Relations, Heinz USA, 23 Oct. 1986. and Marie McDermott, Manager, Thomas J. Lipton, 5 November, 1986, quoted in Webb, Lang and Tucker, *Food Irradiation — Who Wants It*? Thorsons, USA, 1987

41 Preliminary results published in Webb, Lang and Tucker, *Food Irradiation — Who Wants It*? Thorsons USA 1987. Latest survey being undertaken by Food and Water Inc, New York, 1989

42 Statement from Coles-Meyer to Tony Webb and Beverley Sutherland-Smith, reported in *Food Irradiation in Australia*? NCSFI, Melbourne, 1988

43 As ref. 7

44 Consumer Interpol Memo. International Organization of Consumers Unions (IOCU), Penang, Malaysia, August 1986

45 James Dietch, 'Economics of Food Irradiation'. *Food Science and Nutrition*, 17: 307, 1982

46 *Food Irradiation Processing*. Proceedings of Symposium held 4 March 1985, Washington DC, IAEA, Vienna, 1985
47 *The Financial Times*, London, 7 January 1985. *The Daily Telegraph*, London, 7 January 1985
48 Isotron plc. Offer for Sale by Tender of 3,290,088 Ordinary Shares of 25p each at Minimum Tender Price of 120p per Share. J. Henry Schroder Wagg and Co, 27 June 1985
49 Early Day Motion, *Irradiated Foods: Conflict of Interest*. House of Commons Order Paper, 9 April 1986
50 *Wall Street Journal*, 13 May 1983, 5 July 1983, 6 July 1983, 18 Aug 1983, 19 March 1984
DJ News, 7th Oct. 1983, 29 Dec. 1983, 20 April 1984, 6 June 1984, 22 June 1984, 25 June 1984, 30 July 1984, 17 Oct. 1984, 9 Nov. 1984, 12 Nov. 1984, 30 Nov. 1984, 2 Jan. 1985, 17 July 1985, 17 Oct. 1985, 20 Jan. 1986
51 'Investor Traded in 6 Issues Before News Appeared in Wall Street Journal', *Wall Street Journal*, 6 April 1984
'The Wall Street Journal Has Reported Some Details of M. David Clarke's Share Dealing', *International Herald Tribune*, 7 April 1984
52 Report for IOCU on the Meeting of the Joint IAEA/WHO/FAO International Consultative Group on Food Irradiation Held at Cadarache, France, 18-22 April 1988, prepared by Tim Lang, of the London Food Commission, UK, who attended the meeting as an observer at the request of IOCU, April 1988
53 R. L. Lake, Chief of Regulation for the FDA Bureau of Foods, Washington DC, 1983
54 See Ken Terr, 'The Plutonium Connection — Why is the DOE for Food Irradiation?' *The Nation*, 7 February 1987
55 M. C. Lagunas-Solar and S. M. Matthews, 'Comparative Economic Factors on the Use of Ionizing or Electrical Sources for Food Processing with Ionizing Radiation'. *Radiation Physics and Chemistry*, 25 (1-3), 1985
56 As ref. 7
57 T. Webb, Report to IOCU on RPFI Phase III meeting, Bangkok, Thailand, October 1988
58 G. Van Dijk and T. Tijdink, 'Developments and Issues Relating to Food Irradiation in Europe', *Forecasting and Assessment in Science and Technology*, No. 134, Directorate-General for Science, Researching and Development, Commission of the European Communities, January 1987
59 As ref. 58
60 Survey undertaken for the Australian Consumers Association by Sandra Heilpern in 1986/7
61 Survey conducted by the UK Consumers' Association, Nov. 1986
Food Irradiation Omnibus Research by Marplan Ltd, conducted for the London Food Commission, Jan. 1987, British Market Research Bureau, 1989
62 Titlebaum, Dubin and Doyle, 'Will Consumers Accept Irradiated

Foods?' *Journal of Food Safety* 5: 219-228, 1983
63 Bruhn, Schultz and Sommer, 'Attitude Change Toward Food Irradiation Among Conventional and Alternative Consumers'. *Food Technology*, Jan 1986
64 Brand Group; *Irradiated Seafood Products: A position paper for the seafood industry*. Prepapred for the National Marine Fisheries Service, Washington DC, Jan. 1986
65 Denise Rennie et al., Presentation to seminar on food irradiation Salford University, Manchester, 1986
66 Survey conducted by the UK Consumers' Association, London, Nov. 1986
67 *Food Irradiation*. Omnibus Research by Marplan Ltd, conducted for the London Food Commission, Jan. 1987

CHAPTER 7

1 I. Cole-Hamilton and T. Lang, *Tightening Belts*. London Food Commission, London, 1988
2 World Health Organization, *Document on Food Irradiation*. Adopted by the FAO/IAEA/WHO/ITC-UNCTAD/GATT International Conference on the Acceptance Control of and Trade in Irradiated Food, Geneva, 16 December 1988
Acceptance Control of and Trade in Irradiated Food. Conference proceedings, IAEA, Vienna, 1989
3 The London Food Commission, *Food Adulteration and How to Beat it*. Unwin Hyman, London, 1988
4 Correspondence between London Food Commission and HRH The Prince of Wales, 6 November and 14 December 1987, enclosing copies of documents obtained under the US Freedom of Information Act by the Health and Energy Institute, Washington DC, including: J. W. McCutcheon, Meat Poultry and Inspection Services, US Dept of Agriculture to Frank Olsen, acting for Radiation Technology, 28 Oct. 1986; G. L. McCowin, FDA, to Andrew Welt, Radiation Technology Inc and Andrew Welt to G. McCowin, 17 July 1986; M. V. Bensen, Operation Raleigh to Martin Welt, Radiation Technology, 7 July 1986
5 *Declaration on Food Irradiation* of the Asia-Pacific Regional Conference on Food Irradiation, Canberra, Australia, 9-11 November 1988
6 Correspondence from Project for Ecological Recovery and 24 consumer, environmental and other non-governmental organizations to the Prime Minister of Canada, 14 Sept 1988; Correspondence from Probe International and 30 Canadian organizations to the Prime Minister of Thailand, 13 September 1988
7 R. Collard, *Total Quality: Success through People*. Institute of Personnel Management, 1989

Index